Lithium and Cell Physiology

Ricardo O. Bach   Vincent S. Gallicchio

Editors

Lithium and Cell Physiology

With 12 Illustrations

Springer-Verlag
New York Berlin Heidelberg
London Paris Tokyo Hong Kong

Ricardo O. Bach
President, Bach Associates, Inc., Lake Wylie, South Carolina 29710, USA, and
formerly Manager, Research Department, Lithium Corporation of America, Bessemer
City, North Carolina 28016, USA

Vincent S. Gallicchio
Associate Professor, Division of Hematology/Oncology, Department of Medicine,
Department of Microbiology and Immunology, and
Department of Clinical Sciences, University of Kentucky Medical Center, Lexington,
Kentucky 40536, USA

Library of Congress Cataloging-in-Publication Data
Lithium and cell physiology / Ricardo O. Bach, Vincent S. Gallicchio,
    editors.
       p. cm.
    Includes bibliographical references.
    ISBN 0-387-97128-9 (alk. paper)
    1. Lithium—Physiological effect. 2. Cell physiology. I. Bach,
Ricardo O., 1917–   . II. Gallicchio, Vincent S.
    [DNLM: 1. Cells—physiology. 2. Lithium—metabolism. QV 77.9
L77411]
    QP535.L5L57 1990
    612'.01524—dc20
    DNLM/DLC
                                                        89-21751

Printed on acid-free paper.

Typeset by Publishers Service, Bozeman, Montana.
Printed and bound by Edwards Brothers, Ann Arbor, Michigan.
Printed in the United States of America.

9 8 7 6 5 4 3 2 1

ISBN 0-387-97128-9 Springer-Verlag New York Berlin Heidelberg
ISBN 3-540-97128-9 Springer-Verlag Berlin Heidelberg New York

Preface

Bioinorganic Chemistry .... has been defined as the Science which brings Inorganic Chemistry into Life, by recognizing the fundamental Importance of Life Dependence on Metal Ions.—Antonio V. Xavier, Lisbon, October 1985 (Preface to "Bioinorganic Chemistry," A.V. Xavier, Ed., VCH Verlag, 1986)

Recent advances in lithium usage in the biochemical, pharmacological, physiological, and clinical fields have produced a plethora of new information. It became evident that the time was propitious to develop a text to describe the role of lithium in its multiple physiological manifestations. Springer-Verlag New York Inc., encouraged us to proceed with this task and we are indebted for its sustained support of this project. We are also very fortunate to have secured contributions from many scientists and clinicians who are noted experts in their respective areas. It is their dedication and effort, measured by the quality of their respective chapters, that will be the test of success for this book. We hope, too, that this work will stimulate the formation of novel ideas and new research pursuits related to the role of metal ions in the basic mechanisms of life.

Ricardo O. Bach

Vincent S. Gallicchio

Contents

Contributors

CHARLES E. ANDERSON, PH.D.
Professor, Department of Botany, North Carolina State University, Raleigh, North Carolina 27695-7612, USA

RICARDO O. BACH, PH.D.
President, Bach Associates, Inc., Lake Wylie, South Carolina 29710, USA

ROBERT H. BELMAKER, M.D.
Professor of Psychiatry, Beer Sheva Mental Health Center, Ben Gurion University of the Negev, Beer Sheva, Israel

VINCENT S. GALLICCHIO, PH.D. MT (ASCP)
Associate Professor, Division of Hematology/Oncology, Department of Medicine, Department of Microbiology and Immunology, and Department of Clinical Sciences, University of Kentucky Medical Center, Lexington, Kentucky 40536, USA

ARNE GEISLER, PH.D.
Associate Professor, Department of Pharmacology, University of Copenhagen, Copenhagen, Denmark

YORAM GIVANT, PH.D.
Department of Psychiatry, Jerusalem Mental Health Center, Jerusalem, Israel

DAVID A. HART, PH.D.
Professor, Department of Microbiology and Infectious Diseases, Faculty of Medicine, University of Calgary, Health Sciences Center, Calgary, Alberta, Canada

DAVID F. HORROBIN, M.A., PH.D., D. PHIL., B.M., B.CH.
Research Director, Efamol Research Institute, Kentville, Nova Scotia, Canada

JAMES W. JEFFERSON, M.D.
Professor of Psychiatry, Director, Center for Affective Disorders; Co-Director, Lithium Information Center, University of Wisconsin Hospital and Clinics, Madison, Wisconsin 53792, USA

ZEV KAPLAN, M.D.
Department of Psychiatry, Ben Gurion University, Beer Sheva, Israel

GERALD LANCZ, PH.D.
Associate Professor, Department of Medical Microbiology and Immunology, University of South Florida College of Medicine, Tampa, Florida 33612-4799, USA

PESACH LICHTENBERG, M.D.
Department of Psychiatry, Jerusalem Mental Health Center, Jerusalem, Israel

ARNE MØRK, CAND. SCIENT.
Department of Pharmacology, University of Copenhagen, Copenhagen, Denmark

C. IAN RAGAN, M.A., PH.D.
Director of Biochemistry, Neuroscience Research Center, Merck, Sharp and Dohme Research Laboratories, Harlow, United Kingdom

GABRIEL SCHREIBER, M.D., PH.D.
Department of Psychiatry, Ben Gurion University, Beer Sheva, Israel

SOFIA SCHREIBER-AVISSAR, PH.D.
Department of Pharmacology, Ben Gurion University, Beer Sheva, Israel

STEVEN SPECTER, PH.D.
Associate Professor, Department of Medical Microbiology and Immunology, University of South Florida College of Medicine, Tampa, Florida 33612-4799, USA

ELISA G. TRIFFLEMAN, M.D.
Resident, Department of Psychiatry, University of Wisconsin Hospital and Clinics, Madison, Wisconsin 53792, USA

JOSEPH ZOHAR, M.D.
Department of Psychiatry, Ben Gurion University, Beer Sheva, Israel

1
Some Aspects of Lithium in Living Systems

Ricardo O. Bach

Is Lithium an Essential Trace Element?

The "natural" presence of lithium (Li) in the crust of the earth has been estimated to be in the order of magnitude of 65 grams/metric ton (10 eq/mt). In sea water the concentration is around 0.2 gram/metric ton (30 meq/mt or 0.03 meq/liter). The natural presence in animals or humans had not been determined reliably until recently, but one must assume that Li is present as a trace. An element of similar distribution on the earth's surface such as Zn indeed is found in small discrete quantities in tissues and one knows a fair amount of its metabolic functions. Another, even more extreme example is cobalt, which is found in seawater at a concentration of 0.0005 gram/metric ton or 1/400 of the amount of Li. Nevertheless, its function as the central ion in the Corrin ring system of coenzyme B12 is considered as essential. Vitamin B12 is required by human beings as one of the most potent biological compounds known, in an amount of 0.0001 mg/day, which amounts to 0.000004 mg/day of cobalt. Concentration of B12 in the human body is so small that cobalt has never been classified as an essential trace element. It probably must be considered as such.

By simple parallel argument, one must assume that Li is indeed present in extremely small quantities in animal and human tissues. It really is of little importance whether it is or is not classified as a trace element. The real significance might be the question of its essential or not so essential metabolic functions. A compound such as cobalamine defines the role of its inorganic ion with strong liganding bonds to a specific organic moiety (the Corrin ring system) as "essential." The same unique bonding cannot be found for the Li ion. If Li is present in the human body, it is distributed all across its multiple constituents. Indeed there must be a "bonding system" but it should resemble in all likelihood the type of ionic interaction found for related macroconstituent ions such as Na, K, Mg, and Ca. Indeed we can expect nothing unique in this bonding mechanism: while it resembles that of the other alkaline and alkaline earth metal ions in quality, only the quantitative aspects are suspect to be different. This would have to do with the size of the hydrated ion and with location and configuration of receptive sites in the cellular environment.

The natural level of lithium in the blood of healthy individuals has been measured (1). The value in the plasma was around 90 nano equiv/liter and in the erythrocytes around 95 nequ/liter. Thus the distribution is approximately at unity. The doses to be established for the treatment of MDP (Manic Depressive Psychosis) are about four orders of magnitude higher in the serum and about 3 to 4 orders higher in the erythrocytes. Studies emanating from India showed average serum lithium levels of 30 nequ/liter, with a variation from 14 to 72 nequ/liter (2). A more recent study with vastly improved analytical techniques reported values of 1.8 to 4.4 nanograms/gram of dry weight (3). Assuming that the dry weight is about 5% to 6% of the total, this is an average value of about 70 nequ/liter in whole blood, a value which coincides with the one reported by the Russian authors.

These same authors later investigated the effect of the Li^+ content in the drinking water in different areas of the Soviet Union (4). There is a direct correlation between the Li^+ in blood with the Li^+ content of the water supply. The values of the 1980 studies were from Leningrad, where the lithium content of the water supply is very low. The cities of Khar'kov and Erevan have very high Li^+ in water values, around one order of magnitude higher than at Leningrad. The measured values in plasma in Leningrad, Khar'kov, and Erevan are 110, 700, and 750 neq/liter respectively; in erythrocytes 60, 310, and 290 neq/liter respectively. The ratio is much lower than unity—about 0.5. This contradicts Fleishman's 1980 data and coincides with what we know of this ratio from patients under lithium treatment.

In a recent study (5) in the Philadelphia, PA area serum lithium values were determined on 54 healthy volunteers. The results, obtained by an advanced AA technique, fall in the same order of magnitude as the Leningrad results obtained by Fleishman (4), which were based on an Isotope Dilution Mass Spectrometry method.

Clarke et al. (3) consider the narrow range for lithium in blood as pointing to a strong indication that lithium is an essential element. Schou (6) in turn argues that the 3 to 4 order of magnitude difference between the natural level of lithium and the one required for a successful treatment of manic depressive illness precludes the argument of essentiality. He is joined in this opinion by various other researchers (7, 8). The authors (8) make the interesting statement that: "Lithium has no known function in the human body." This statement is contrasted by Nielsen, who, based on studies by Anke and Pickett, believes it quite likely that lithium indeed is essential (9). The studies by Anke (10, 11) give convincing evidence of the essentiality in goats. A lithium-poor diet lead to pathological conditions such as a marked slower growth rate, a reduced rate of insemination, a reduced rate of conception per nanny, a much higher rate of abortions, and a substantially higher incidence of mortality. Nielsen sums this up in the statement that lithium must have an essential endocrine function in view of all these manifestations. Szilagyi et al. (12) investigate enzyme function, MAO activity, and serotonin turnover in lithium-deprived goats, and find these parameters lower in most cases. Picket (13) finds the evidence for essentiality of lithium in the rat by comparing the vital signs of rats with a normal diet versus those of rats on a lithium-deficient diet.

One can be tempted to extend these findings to humans, but there is no direct experimental evidence on hand which takes all factors into account: the drinking water as well as nutrition. The well-known studies by Dawson (admission rate differences to mental hospitals) and Voors (decreased incidence of atherosclerosis) deal exclusively with the content of Li in the drinking water (14, 15). However, taking into account these indications in conjunction with the data on goats and rats, a very strong case must be made for the essentiality of lithium. This in addition deserves credence from the evolutionary point of view, which was aptly introduced by Horrobin and Lieb (16).

The essentiality of the lithium ion in plants presents an intriguing question, which is treated in detail in Chapter 3 of this book. On the one hand, as stated, lithium is not required for a plant to complete its life cycle, however numerous instances of enhancement of growth by small quantities of Li^+ added to the environment have been observed. It is remarkable that this enhancement can be achieved by very small quantities to be followed with the beginning of toxic manifestations (chlorosis) when additional lithium is added to the growth medium. Indeed there seems to be a very narrow range of beneficial action. This resembles the delicate equilibrium between ineffective low doses of lithium carbonate and excessive doses, which show toxic side effects in the treatment of human manic psychosis. Thus no direct evidence for the essentiality of lithium in plants is on hand. However the statement that in plants lithium is not an essential element may be somewhat premature and deserves to be investigated by experimental means.

Doses of Lithium Ion In Vitro and In Vivo

Manic/depressive patients are treated with lithium effectively when the serum concentration of Li^+ is between 0.5 and 1.0 mM/L. Daily administration of 600 to 900 mg/day of Li_2CO_3 (20 to 30 mM of Li^+) usually achieves this goal. The average human serum volume is about 6 liters, thus there are between 3 and 6 mM of Li^+ contained in the serum, or about 20% of the daily intake. The serum weight is about 10% of the weight of the human body. Thus on the average the concentration in the remainder (or 90%) of the human body is about 0.2 to 0.5 mM of Li^+. If we consider the well-known difficulties of Li^+ to traverse certain membranes, where external versus internal equilibrium concentrations can be as high as 3:1, it seems logical to assume that intracellular concentrations in patients on Li therapy can be as low as 0.1 mM or less. The lithium ratio of around 0.1 to 0.2 is well known. This is the ratio of Li^+ in erythrocytes over Li^+ in the serum (18).

The Effect of Certain Membranes

Even lower concentrations can be assumed in the adrenergic centers in the brain, because of the impediment the blood-brain barrier imposes on electrolyte transport. The Li^+ concentration effective in a normal neural microenvironment may be quite small indeed. Could it be that it is correlated to the evolutionary "natural" presence of the element on earth?

Consequences of the Lithium Ratio

A slight change in electrolyte concentrations in this environment could prepare for significant pathological consequences. To reestablish a microenvironment where Li^+ returns to its "normal" level, a very large load of Li^+ is necessary in the serum. It could be that several orders of magnitude higher concentrations are demanded to obtain this objective. This is why loads of up to 1 mM/L in serum must be maintained. Regrettably this concentration is approaching toxic levels.

Where Is the Li^+ Located?

This toxicity can also be explained by the fact that the distribution of Li^+ throughout the human body is widely varied. It is very likely that there are areas in the body where Li^+ is concentrated substantially. Such areas are likely to be the triphosphate moiety of the nucleic acids, the phosphoinositol and other polyphosphatidyl systems. Enzyme complexes located in the protein phase of the proteolipid membranes with their metal ion cofactor specificity are other areas where Li^+ may accumulate. Last but not least, Li^+ is resorbed by the bone mass, some of it in an irreversible fashion. This leads us to another unsolved problem.

Specific Pharmacokinetics: Administration Versus Clearance

One might assume that all lithium administered is eventually cleared by urinary, fecal, and perspiratory elimination. Some data exist that indicate elimination in the high 90%. If one assumes that Li is an essential element in the evolutionary context, the amounts retained by bipolar patients may correlate to a lack of very small amounts of Li, which might be the cause of these pathological conditions. The analytical data of Li^+-in and Li^+-out are much too coarse to find short time differentials, verifying or denying this theory.

Lithium Concentration of In Vitro Antiviral Tests

Skinner et al. (19) and Specter and Bach (20, 21) have demonstrated that Li^+ effectively interferes with the replication of DNA viruses in vitro. The concentration at which the interference becomes noticeable in the relatively short time span (up to 10 days) is above 10 mM/L, one order of magnitude higher than the maximum serum concentration possible before toxic side effects become noticeable. A concentration of 10 mM/L in the serum would be lethal!

Does This "High" Concentration Invalidate the Antiviral Tests?

The question arises whether such concentrations of Li^+ could be of meaningful therapeutic value. There are answers:

The in vitro environment is a cellular one. The toxicity of Li^+ was established to be at around 50 mM/L. This means that the Li^+ in the aqueous humeral part must be in a range up to those levels in order to achieve penetration into cellular organelles across multiple barrier membranes. The virions seem to be more vulnerable than the medium cells per se, which are much more tolerant of these relatively high Li^+ concentrations.

Bach and Specter (22) investigated some kinetic aspects of such in vitro antiviral action. Instead of creating ab initio a certain (high) level of Li^+ in the culture, smaller amounts were added daily. The daily amount per se in a one time addition would have no or very little antiviral effect. However the repeated addition was substantially more effective than the same amount in one shot. Also, while there was never any cytotoxicity observed, the daily addition summed up to 100 mM/L of Li^+, a level which in a one time addition would obliterate the cell culture. The meaning of this, while not totally explicable, could be the uneven distribution of Li^+. Retention of Li^+ in specific sites (see earlier), where it not only would be innocuous to the microenvironment but also most likely would perform its antiviral function. Let us not forget that in the psychotherapeutic treatment daily addition of substantial doses are required also.

Lieb reported the cessation of herpetic lesions in manic/depressive patients on lithium (23). A severe case of VZV facialis was treated by oral administration of lithium carbonate, created serum levels of 0.6 to 0.9 mM/L of Li^+. It was very successful (24). It is inescapable that this serum level has its corollary in the 20 or more times higher concentration of the in vitro tests.

Ionophores

The transport of ions across biological membranes is of primary importance in most metabolic processes. In the following, literature is reviewed which deals primarily with the transport of the lithium ion. This is of tremendous interest to all who are concerned with the pharmacokinetics of lithium, mostly to psychiatrists, but also to researchers concerned with the hematopoietic, immune-enhancing, and antiviral properties of the lithium ion. It is quite apparent that in medical practice, in order to achieve therapeutic action, the bodily systems are usually loaded close to toxic levels. It is also known that lithium does not easily transgress the blood-brain barrier and that the concentrations needed to bring the adrenergic/cholinergic systems into equilibrium resembling "normalcy" are one order of magnitude lower than the one created in the serum. If special vehicles could be designed that carry the lithium ion to the sites of its designed action, one would not have to overload the whole intracellular and intercellular systems with lithium near toxic levels.

Another purpose of this review is to show that there may be agents with preferential ties to lithium. This in turn may serve as a means of separation from other ions, which are usually associated with lithium in ores and brines. Thus, raw materials that are difficult to exploit for their lithium values alone could become economically accessible. The almost inexhaustible amount of lithium in the oceans thus could be made available, to cite the most obvious example.

Pacey (25) stated in 1983: "Detailed investigations into the lithium selectivity of macrocyclic compounds are still in their early infancy." In the intervening two-three year period a lot has been learned about the factors leading to selectivity for Li, Na, K, and other ions. Most astonishing is the fact that cyclicity is *not* a requirement for strong liganding of these ions.

Researchers at Osaka University were foremost in the development of lithium-specific crown ethers and spherands, mostly for analytical purposes. Kimura et al. (26) describe lipophilic 14-crown-4 derivatives containing nitrophenols as chromophores. The absorbance of the Li complex is at 500 nm versus that of K, Na, Rb, and Cs which has its maximum at around 370 nm. The same authors describe how the lithium is transported selectively through polyethylene membranes containing the crown compounds. Especially relevant is a recent paper by Misumi and Kaneda (27). These authors successfully attempt a classification of the different types of "oversize" ring systems with respect to their tendency to form Li complexes. Dinitrophenyl-azaphenols contained within a polyether crown or a spherand containing protonic substituents inside the ring systems "have unique structural features to exhibit lithium ion selective or specific coloration by nesting type complexation."

Most relevant however are some very recent publications by Suzuki and Shirai (28) and Hiratani and Taguchi (29). Both publications describe strong interactions of the lithium ion with open chain structures. The first time such affinity has been described was by Shanzen, Samuel, and Korenstein (30) in compounds with the noncyclic dioxadiamide structure. A short discussion of the possible reasons for lithium selectivity by the O-N-O-O structures, early proposed by Birch, is found in (31). Suzuki et al. describe natural carboxylic polyethers as lithium ionophores. Two compounds show outstanding lithium selectivity factors: (1) Macrocyclic monensin and its monoacetate (an antiprotozoal agent) and (2) ionomycin methyl ester (an antibiotic). Hiratani and Taguchi synthesized noncyclic polyether dicarboxylic acids. These were placed in nonpolar organic solvents and the transport of the corresponding alkali ion complexes was observed. The rate of transport revealed a strong preference for the Li^+. The specificity is explained by the folding of the polyether chain around the naked lithium ion. This may result in a pseudocyclic complex structure somewhat reminiscent of the induced-fit property of enzyme structure to harbor metallic cofactors.

For quite some time the theoretical factors for the binding energies of the different alkali metal ions by cryptands, crown ethers, spherands in general, and other cyclic organizations have been discussed. The first and most obvious factor is the correlation of ionic size versus the cavity of the host. By and large it proved to be a fruitful correlation. A recent paper illustrates this principle again (32). Here, 12-crown-3, hexamethyl 12-crown-3, and 16-crown-4 were compared in their complex formation with different alkali cations. The latter showed high selectivity for Li over the other cations, but the binding energies were lower than hydration energies and therefore lack interest. The permethylated 12-crown-3 and 12-crown-3 itself show preference for Li over the other alkali ions, but only the hexamethyl crown shows practical applicability and high selectivity against

Mg^{++} for instance. This indeed is very important. It appears that the preponderance of "axial" methyl groups provides a shield against the higher coordination demanding ions, such as Na, K, Cs, and Mg. Much higher binding energies are, of course, noted in cryptands, as noted by the original research efforts at Strasbourg by Jean Marie Lehn (recent Nobelist, together with Donald Cram and Pedersen). The "lithium connection" is shown in (33): di-, tri-, and tetra-oxa-diaza-bicyclo-icosanes all form strong Li complexes. Structural analysis and stepwise solvolysis are used to compare the Li versus Na complexation energies.

Another example of open versus cyclic structures with respect to preferential complexation is found in (34). Macrocyclic diamides of tetramethyl-dioxaoctanedioic acid show a highly preferential logK for Li over other alkali and alkaline earth cations including $NH4^+$ and H^+. The structure is an eleven member ring system, with the typical O-O-N-O configuration. It also has the shielding effect of "axial" methyl groups. This preferential complexation was observed also when the ionophore was incorporated in a pvc based membrane. Actual concentration potentials could be measured across the membrane. The authors compare this with the very similar noncyclic dioxadiamides (30), where the specific Li effect is much less pronounced.

An exhaustive investigation of "host-guest complexation" was carried out by Donald J. Cram (Nobelist, see earlier) and co-workers (35, 36). First an attempt is made to classify further the different ringstructures. The crown compounds of Pedersen (co-Nobelist with Lehn and Cram) are renamed chorands. The cryptands (Lehn) are three-dimensional due to trisubstituted N, which leads to three ring systems. A ring system where benzene rings are linked to each other in the meta position is called a spherand. It is easy to see that combinations of these sytems can be synthesized. A combination of a chorand (crown) and a spherand is called a "hemispherand" and one of a cryptand and a spherand is a "crypta-hemispherand." The highest order of affinity is found in the spherands for all ions and specifically lithium. Cram did extensive work on structural aspects. The spherands and hemispherands of great promise contain methoxy groups on the benzene moieties. It was found that all free base electron pairs were turned inward when a metal ion entered the cavity.

There are now a substantial number of lithium-specific macrocyclic compounds described in the literature. The synthetic methods and the raw materials are not exotic and the availability of these host compounds at reasonable prices would only be a consequence of sufficient demand.

Hruska (34) shows how a rather simple eleven-member ring system when incorporated in polyvinylchloride becomes a lithium-specific membrane. Some acres of such a membrane, which separates seawater and pure water, should see lithium ions almost exclusively permeate into water in sufficient quantity to make exploitation possible. Misumi's (27) compounds could possibly be used the same way.

These same hosts, most likely, would not be toxic at the proper concentrations. Research should concentrate on concurrently administering lithium-specific hosts with modest amounts of lithium compounds, much below the toxic levels.

If these lipophilic hosts carry the lithium across the membranes into the proper cellular environment, the serum and other humeral systems should not become overloaded with lithium ion.

A connection between these macrocyclics and noncyclic natural carboxylic polyethers must exist because some of these exhibit lithium selectivity. They belong to compounds known for therapeutic properties, antibiotic and antiprotozoal. Synthetic efforts show that linear compounds can fold in such a way, to form hostlike lithium-specific cages. The rules of molecular geometry are being investigated by Donald Cram for these macrocyclic structures.

Is it accidental that the most notable researchers of these cyclic "wonders," Cram, Pedersen, and Lehn, received the Nobel prize for their work? I believe not. Not one of them originally came from the biological side, but all of them must have become keenly aware of the biological significance of these macrocyclics and their kinship with the protein-based enzyme systems. The evidence of metal ion specific "niches" or configurations in these proteins is a well-known and extensively described fact, and can be related to functions of most enzyme systems.

Lithium and the Prostaglandin

Horrobin (37) makes the case of "paradoxical" actions of lithium in an elegant way. Inherent in his proposition is the assumption of the presence of the lithium ion in all the systems he discusses. The question is that of either excessive or deficient quantity. The main concern is with the immune systems, especially their link with the prostaglandins. The biosynthetic path leads from essential fatty acids such as linolenic acid to di homo gamma linolenic acid (DGLA) which then converts to prostaglandins by mediation of a cyclooxygenase enzyme complex. The Li ion in therapeutical doses shows a distinct blocking of the cyclooxydation of free DGLA to prostaglandin E1. It is known that manic patients have high levels of PGE1, thus the action of Li can be antimanic. The manic phase, in bipolar illness, alternates with the depressive phase in a flip/flop fashion, because the PGE1 levels in the depressed state are abnormally low. It is easy to see why Li does not directly address the patient when in the depressed state. It is well known now, however, that Li will indeed provide protection against depressive lapses, as well as against manic episodes. But only with mania is there an immediate therapeutic effect. In the same way, excess PGE1 leads to excess suppressor T-lymphocyte, which is usually connected with some impairment of immune functions. The reduction of PGE1 production from DGLA via Li reestablishes some part of immunity, with the proper equilibrium between helper and suppressor T-lymphocyte titers. In contrast to quasi-normal equilibrium, the PGE1 concentration can sink to abnormally low levels, as seen in depression, and may lead to autoimmunity with consequences such as affliction with rheumatoid arthritis.

In a more recent article (38) further insight is gained into the metabolism of essential fatty acids, prostaglandin and other products of the arachidonic acid cascade. It seems that Li has no effect on the prostacyclin/PGF relationship, whereas it shows some interference with the formation of thromboxanes. Horrobin assumes that lithium interacts with PXA2 synthetase. In this new set of tests lithium shows again no action on PGE2, which contains one hydroxygroup more than PGE1. The latter is again strongly reduced under the influence of the lithium ion.

At very high concentrations of Li^+ the outflow of all fatty acids as well as all the previously mentioned products of the arachidonic acid cascade are strongly reduced. This effect parallels the mode of action of corticosteroids which are used externally for their anti-inflammatory properties. The reasonably well-known use of lithium in herpes simplex I and II infections and varicela zoster based shingles could be related to the earlier described reductions of PGE and thromboxane (39, 40). One might assume that the lipoxygenase catalized arachidonic cascade leading to HETE and leukotrienes could also be inhibited by the Li^+ to some degree. This could be an important factor in the treatment of rheumatoid arthritis.

Unlike aspirin and other nonsteroidal anti-inflammatory analgesics which mainly act on the cyclooxygenase by chemical transformations (acetylation in the case of aspirin), lithium does not transform the enzyme system other than to act on its metallic cofactor, prevalently the Mg^{++}. Therefore the cyclooxygenase system is only partially inhibited when Li concentrations become high enough to replace some of the Mg^{++}. When the pendulum swings back in favor of higher Mg concentration, the enzyme activity increases. This is why Li is effective in bipolar disease and should not pose a threat to autoimmune reactions. It also would follow that Li might not be effective against rheumatoid arthritis and other inflammatory manifestations. However, as seen earlier, other related mechanisms might point to the opposite.

Membranes

Homeostasis

In the case of bipolar illness, the contrast is "bipolar health," characterized by mood swings rather than by flip/flops. An equilibrium of helper and suppressor functions means normal immune response capabilities as well as bodily resources to ward off autoimmunities. The natural presence of trace amounts of lithium in the nanoequivalent/liter range may be one of the guarantors of this homeostasis. If the equilibrium is disturbed by external factors such as microbial and viral infections, manipulation of the lithium level in cellular and humeral environs may be a remedial option. We are very far from understanding how to handle this potential tool. Certainly this can only be successful in the context of establishing proper ratios between the macroconstituents Ca, Mg, Na, K, Cl, P, S, and Li. In addition one cannot dismiss potential interactions with essential trace elements, such as Fe, Co, Zn, Cr, Mo, and Se, to name just a few. Transport phenomena across phospholipid membranes have been studied for Li and were found to be regulated by Na, K, and Ca concentrations on one or the other side of the membrane. It is interesting to note that the driving force established in inorganic systems by the voltage across the membranes of concentration cells bears little resemblance to the transport phenomena in biological systems, where we often observe "uphill, against the gradient" migration.

How the Li⁺ Ion Permeates the Phospholipid Membranes

Pandey (41, 42) described the different modes of transportation of Li⁺ across the RBC membrane. Four different pathways are recognized: leak (which is similar to diffusion), Li⁺/Na⁺ exchange, anion exchange ($LiCO3_-$), and Na⁺–K⁺ pump (Na/K ATPase mediated). The steady state between red cells and plasma is defined as the Li ratio, and is usually lower than unity. The concentration of Na⁺ determines in which direction Li⁺ flows across the membranes: the experiments demonstrate an uphill extrusion of Li⁺ from red blood cells which is driven by a Na⁺ gradient which mandates exchange of Na⁺ in the opposite direction.

Enters the Magnesium Ion Cofactor

A significant experiment describes the only Li ratio higher than unity. When lithium-loaded cells at 1.0 meq/L are incubated in a substrate of high Mg^{++} and 1.5 meq/L of Li⁺, the concentration of Li⁺ in the RBC's ascends to close to 2 meq/L of Li⁺ after four hours. It appears that Mg^{++} promotes this action and adds its osmotic pressure to that of the Li⁺. The similarity of the ionic character of these ions is again demonstrated in a biological environment. It is also noted that this "against the gradient" migration of Li⁺ is inhibited by ouabain (strophanthin), a cardiac glycoside. One might assume that the presence of ouabain leads to an interaction with Mg^{++} rendering it impotent to promote the "against the gradient" migration of Li⁺. Ouabain has multiple OH groups, which can coordinate with the highly polar Mg^{++} easier than with Li⁺. The enzyme Na⁺/K⁺–ATPase, which is Mg^{++} activated, also is inhibited by ouabain.

Lithium/Magnesium Competition

PGE1 is stimulating (43) the activity of adenylate cyclase, an enzyme which has Mg^{++} as the metallic cofactor. Li⁺ has shown a decrease in activity of formation of cAMP in this system. It is interesting to note that an increase of Mg^{++} in the substrate decreases the inhibiting action of Li⁺. It seems that the lithium ion partially replaces the Mg^{++} leading to some inhibition. Upon increasing the concentration of Mg^{++}, the system is reactivated. This is another instance of indirect confirmation of the chelation theory.

Another Look at Membranes

The mode by which one ion can replace another totally or partially is defined by the ionic equilibrium laws for nonbiological aqueous systems. In biological systems these laws help to give approximate estimates of shifts in equilibrium, but they will never be able to give accurate quantitative descriptions. The very nature of permeable membranes across which the ions must migrate influences the direction and course of this migration. There are ion-specific sites along the channels which determine the conformation when ions are in the process of migration. Indeed it has been demonstrated that there are conformational changes associated with the permeation of different ions (44). A Ca binding channel was exposed to permeations of Cs, K, and Na ions. Changes in the pH reveal that the binding destabilizes the protonated state of the external channel surface. When the Li^+ migrates into the membrane-bound enzyme systems, which are activated by Mg^{++}, such conformational changes must take place. When viral DNA, as an example, is copied for the replicative process, one can assume that not only the nucleic acid sequence but also the associated proteins suffer conformational changes when the univalent Li ion replaces the divalent Mg ion. Such a change can lead to the interruption of replication. It is however quite clear that the binding preferences for protein conformations can be quite different from those of the nucleic acids, where most of the ionic interplay is located on the triphosphate moiety. In the latter case, ionic equilibrium laws can give reasonably good estimates of degrees of replacement of one ion by another (22, 45).

The conformation of the associated proteins is influenced by a change of the occupying ion, as shown by Hess (44). Both events contribute in some manner to the cessation of viral replication, when Li^+ replaces Mg^{++} as the cofactor. This has been clearly stated by Skinner (19).

The Magnesium-dependent Na,K–ATPase as a lithium transport system (46) is responsible for the normal distribution of K, Na, and Ca across membranes, specifically the neuronal membranes as discussed in this paper. In manic depressive illness a reduced activity of Na,K–ATPase has been noted with the consequence that Na and Ca increase intracellularly. Free Ca ion would be responsible for uncontrolled neurotransmission. When Li enters this system, an activation of the Na,K–ATPase is observed, which leads to substantial egress of Ca and Na, and neuronal transmission will begin to return to more normal modes. Only at very high Li concentration will Li start to replace the Mg as cofactor and will thus inhibit the total enzyme system.

The course of normalization by Li^+ is also manifested by the level of cAMP excretion. The manic phase has elevated urine cAMP, which upon treatment with Li^+ will return toward normal levels. The depressive phase is accompanied by low levels of excreted cAMP, which upon treatment with Li^+ will begin to increase toward normal levels.

Beyond doubt, the picture emerges clearly that ionic equilibria play a very important role in this push-pull situation of bipolar illness. Not only is Li^+ competing for the cofactor position of the enzyme system, it also is modifying the flux of other ions, mono and divalent, such as Na and Ca. The membrane channels are suffering alterations upon the arrival of new ionic wanderers (Hess) and these changes will allow a release of excess Na and Ca ions bottled up within the neuronal cell, where they were creating dysfunctions. This, additionally, could account for the antimanic and antidepressive effects of the lithium ion.

Summary

The essentiality of lithium can be supported not only on the basis of evolutionary considerations, but by other factors as well. There are animal studies which support lithium's role as a sine qua non for physical health in the case of goats and rats. While the direct evidence for the human metabolism is somewhat less clear, there are still strong indications that lithium in the range of nanoequivalents per liter is an essential element.

The application of Li^+ in the treatment of bipolar illness is effective only at high serum concentrations and the antiviral activity in vitro also demands a level many orders of magnitude higher than the ten to a hundred nanoequivalents range, which could be assumed to be essential. It can be assumed that the microenvironments of the viral replicative mechanism and of the mood-controlling neuronal cells may be susceptible to slight variations of the electrolyte composition. In the case of viruses very small amounts of lithium ion seem to interfere with enzymatic processes essential to replication. In the case of bipolar illness one might speculate that a slight deficit or excess versus the essential level is amongst the causes. Supply of additional Li^+ serves as a prompt correction in case of mania and has been proven to be a valid preventive measure against recurring depressions. The actual levels of Li^+ in the microenvironment are likely to be much smaller than the humeral concentration. This most likely is due to the impediments of migration through multiple membranes of the cells and organelles.

The therapeutic levels of lithium in vitro and in vivo in the serum are very close to cytotoxicity and this is the greatest detriment of the lithium treatment. A possible way out of this potentially dangerous situation is the concurrent application of lithium-specific ionophores. They must have lipophilicity and should be non-toxic. The beginning of research into this alternative is on hand. If a successful ionophore is used, it may not be necessary to "load up" the serum to levels close to toxic.

Lithium blocks the cyclooxidation of DGLA to prostaglandin E1. The cyclooxygenase system has Mg^{++} as its cofactor. The likely mechanism of the blockage is the interference by the Li^+ with this cofactor, similar to the antiviral action of the Li^+, which has been shown to be mediated that way. The theory of chelation thus is common to both: blockage of PGE synthesis and antiviral action.

The specific character of phospholipid membranes regulates the permeation of lithium and the other ions such as Ca, Mg, Na, and K. There are now indications that Li^+ may contribute to homeostasis, thereby exerting its therapeutic value. Magnesium ion promotes the penetration of lithium into RBC's, which is another example of the intimate correlation of these two ions. It can be stated unequivocally that an understanding of the many effects of lithium in living systems is only possible in the context of the regulation of the composition of the electrolyte. The chelation theory may serve as a first step toward this goal.

References

1. Fleishman, D.G., Gurevich, Z.P., Solyus, A.A., Baklanova, S.M., & Skul'skii, I.A. (1980). The natural lithium content in the blood of man and certain animals. *Doklady Akademii Nauk SSSR, 254*(6), 1497–1501.
2. Jathar, V.S., Pendharkar, P.R., Pandey, V.K., Raut, S.J., Doongaji, D.R., Bharucha, M.P.E., & Satoskar, R.S. (1980). Manic depressive psychosis in India and the possible role of lithium as a natural prophylactic. II-Lithium content of diet and some biological fluids in Indian subjects. *J. Postgrad. Med., 26*(1), 39–44.
3. Clarke, W.B., Webber, C.E., Koekebakker, M., & Barr, R.D. (1987). Lithium and boron in human blood. *J. Lab. Clin. Med., 109*, 155–158.
4. Fleishman, D.G., Solyus, A.A., Gurevich, Z.P., & Bagrov, Ya.Yu. (1985). Lithium in the human body. *Hum. Physiol., 11*, 430–435.
5. McCarty, J.D., Carter, S.P., Fletcher, M.J., Reape, M.J., & Shindell, S. (May, 1988). Unpublished study by FMC Corporation. (Permission to quote gratefully acknowledged.)
6. Schou, Mogens (1989). *Lithium Treatment of Manic-Depressive Illness*, 4th rev ed, KARGER, Basel-München-Paris-London-New York-New Delhi-Singapore-Tokyo-Sydney, p. 22.
7. Jefferson, J.W., & Greist, J.H. (1987). Lithium in psychiatric therapy. *Encyclopedia of neuroscience*, Vol. 1 (p. 592). Boston: Birkhauser.
8. Calabrese, E.J., Canada, A.T., & Sacco, C. (1985). Trace elements and public health. *Ann. Rev. Public Health, 6*, 140.
9. Nielsen, F.H. (1984). Ultratrace elements in nutrition. *Ann. Rev. Nutr., 4*, 27.
10. Anke, M., Groppel, B., Kronemann, H., & Gruen, M. (1983). Evidence of essentiality of lithium in goats. In M. Anke et al. (Eds.), *Lithium: Spurenelement Symposium* (pp. 58–64). F. Schiller Universitaet, Jena.
11. Anke, M., Groppel, B., & Kronemann, H. (1984). Significance of newer essential trace elements (like Si,Ni,As,Li,V . . .) for the nutrition of man and animal. In P. Bratter & F. Schramel (Eds.), *Trace elements analytical chem. in med. and biol.*, Vol. 3 (pp. 436–440). Berlin/New York: W. de Gruiter.
12. Szilagyi, M., Anke, M., & Szentmihalyi, S. (1983). Influence of lithium deficiency on the enzyme status of goats. In M. Anke et al. (Eds.), *Lithium: Spurenelement Symposium* (pp. 71–75). F. Schiller Universitaet, Jena.
13. Picket, E.E. (1983). Evidence for the essentiality of lithium in the rat. In M. Anke et al. (Eds.), *Lithium: Spurenelement Symposium* (pp. 66–70). F. Schiller Universitaet, Jena.
14. Dawson, E.B., Moore, T.D., & McGanity, W.J. (1972). Relationship of lithium metabolism to mental hospital admission and homicide. *Dis. Nerv. Syst., 33*, 811.
15. Voors, A.W. (1971). Lithium in the drinking water and atherosclerotic heart death: Epidemiological argument for protective effect. *Am. J. Epidemiol., 93*, 259–266.
16. Horrobin, D.F., & Lieb J. (1981). A biochemical basis for the actions of lithium on behaviour and on immunity: Relapsing and remitting disorders of inflammation and immunity such as multiple sclerosis or recurrent herpes as manic depression of the immune system. *Medical Hypotheses, 7*, 891–905.
17. Campbell, A.C., Palmer, M.R., Klionkhammer, G.P., Bowers, T.S., Edmond, J.M., Lawrence, J.R., Casey, J.F., Thompson, G., Humphris, S., Rona, P., & Karson, J.A. (1988). Chemistry of hot springs on the mid-atlantic ridge. *Nature, 335*, 514–519.
18. Beaubernard, P. et al. (1987). Lithium: Clinical study by brain electrical activity mapping. A case report. *Pharmacopsychiat., 20*, 201.

19. Skinner, G.R.B., Hartley, C., Buchan, A., Harper, L., & Gallimore, P. (1980). The effect of lithium chloride on the replication of herpes simplex virus. *Med. Microbiol. Immunol., 168*, 139–148.

20. Specter, S., Bach, R.O., & Green, C. (July, 1986). Inhibition of herpes virus in cell cultures by lithium ions. *Abstr. IXth Inter. Cong. Infect. and Parasit. Dis.* (p. 417). Munich.

21. Specter, S., & Bach, R.O. (1987). Lithium ion induced inhibition of herpes-viruses in cell culture. *Abstr. Ann. Meeting Am. Soc. Microbiol.*, 16.

22. Bach, R.O., & Specter, S. (1988). Antiviral activity of the lithium ion with adjuvant agents. In N.J. Birch (Ed.), *Lithium: Inorganic pharmacology and psychiatric use* (pp. 91–92). Oxford: IRL Press.

23. Lieb, J. (1979). Remission of recurrent herpes infection during therapy with lithium. *N. Engl. J. Med., 301*, 942.

24. Roesli, A., & Bach, R.O. (1984) Unpublished.

25. Pacey, G.E. (1985). Lithium crown ether complexes. In R.O. Bach (Ed.), *Lithium: Current applications in science, medicine and technology* (p. 35). New York: Wiley and Sons.

26. a) Kimura, K., Tanaka, M., Kitazawa, S., & Shono, T. (1985). Highly lithium-selective crown ether dies for extraction photometry. *Chemistry Letters* (pp. 1239–1240). b) Emsley, J. (1985). Smuggling lithium across membrane. *New Scientist, 107*, 23.

27. Misumi, S., & Kaneda, T. (1987). Syntheses of chromophoric ethers and spherands and their selective coloration with lithium ion and amines. *Mem. Inst. Sci. Ind. Res.* (Osaka). *44*, 29–47.

28. Suzuki, K., Tohda, K., Sasakura, H., Inoue, H., Tatsuka, K., & Shirai, T. (1987). Natural carboxylic polyether derivatives as lithium ionophores. *J. Chem. Soc., Chem. Commun.*, 932–934.

29. Hiratani, K., & Taguchi, K., (1987). Synthetic polyether dicarboxylic acids exhibiting lithium ion-selective transport. *Bul. Chem. Soc. Jpn., 60*, 3827–3833.

30. Shanzer, A., Samuel, D., & Korenstein, R. (1983). Lipophilic lithium ion carriers. *J. Am. Chem. Soc., 105*, 3815–3818.

31. Bach, R.O. (1987). Lithium and viruses. *Medical Hypotheses, 23*, 157–170.

32. Dale, J., Eggestad, J., Fredriksen, S.B., & Groth, P. (1987). 1,5,9-Trioxacyclodo-decane and 3,3,7,7,11,11-hexamethyl-1,5,9-trioxacyclododecane: Novel lithium cation complexing agents. *J. Chem. Soc., Chem. Commun.*, 1391–1393.

33. Abou-Hamdan, A., Houslow, A.M., & Lincoln, S.F. (1987). A structural study of the complexation of the lithium ion by the cryptand 4,7,13-trioxa-1.10-diazabi-cyclo(8.5.5.)icosane. *J. Chem. Soc. Dalton Trans.*, 489.

34. Hruska, Z., & Petranek, P. (1987). Macrocyclic diamide based polyvinyl chloride membranes with a high lithium selectivity. *Polymer Bulletin, 17*, 103–106.

35. Cram, D.J., Ho, S.P., Knobler, C.B., Maverick, E., & Trueblood, K.N. (1986). Host-guest complexation. 38. Cryptahemispherands and their complexes. *J. Am. Chem. Soc., 108*, 2989–2998.

36. Cram, D.J., & Ho, S.P. (1986). Host-guest complexation. 39. Cryptahemispherands are highly selective and strongly binding hosts for alkali metal ions. *J. Am. Chem. Soc., 108*, 2998–3005.

37. Horrobin, D.F. (1985). Lithium in the control of herpes virus infections. In R.O. Bach (Ed.), *Lithium: Current applications in science, medicine and technology* (p. 397). New York: Wiley and Sons.

38. Horrobin, D.F., Jenkins, D.K., Mitchell, J., & Manku, M.S. (1988). Lithium effects on essential fatty acid and prostaglandin metabolism. In N.J. Birch (Ed.), Lithium: Inorganic pharmacology and psychiatric use (p. 173). Oxford: IRL Press.
39. Skinner, G.R.B. (1983). Lithium ointment for genital herpes. *Lancet, 288.*
40. Bach, R.O. (1/12/87). Method and composition for controlling viral infections. U.S. Patent Application, Ser. No. 002, 170.
41. Pandey, G.N. et al. (1978). Lithium transport pathways in human red blood cells. *The Journal of General Physiology, 72,* 233–247.
42. Pandey, G.N., & Davis, J.M. (1980). Biology of the lithium ion. In A.H. Rossoff & W.A. Robinson, (Eds.), *Lithium effects on granulopoiesis and immune function* (pp. 16–59). New York: Plenum Press.
43. Wang, Yao-Chun et al. (1974). Effect of lithium on prostaglandine-1-stimulated adenylate cyclase activity of human platelets. *Biochemical Pharmacology, 23,* 845–855.
44. Pietrobon, D., Prod'hom, B., & Hess, P. (1988). Conformational changes associated with ion permeation in L-type calcium channels. *Nature, 333,* 373–376.
45. Birch, N.J. (1976). Possible mechanism for the biological action of lithium. *Nature, 264,* 681.
46. El-Mallakh, Rif S. (1983). The Na,K-ATPase hypothesis for manic depression. II. The mechanism of action of lithium. *Medical Hypotheses, 12,* 269–282.

2
Naturally Occurring Lithium

ELISA G. TRIFFLEMAN AND JAMES W. JEFFERSON

Over the years, many claims and scientific theories have been advanced about lithium, the lightest alkaline metal. Those studies regarding the medical effects of naturally occurring lithium have ranged from the use of lithia spring waters in the treatment of gout to attempts to correlate the incidence of atherosclerotic heart disease with tap water lithium content. Most such studies have suffered from methodological problems, or a lack of replicable or substantive results. Comparisons among studies are made difficult by great variations in measurement techniques and in the widely varying levels of lithium felt to be of critical interest. Yet, these studies raise several questions: Does naturally occurring lithium serve a purpose in human physiology? Is there a lithium deficiency syndrome? Can microgram dosages of lithium be harmful to health? This chapter explores these issues with an emphasis on a critical evaluation of existing data.

Historical View: Lithia Waters

Lithium was discovered in 1817 by Arfwedson in petalite rock (1). By 1848, the ability of lithium carbonate solutions to solubilize urate crystals had been discovered (2). This coincided with a time in western medicine when many ills, including gout, were attributed to the imbalances of uric acid in the blood and elsewhere in the body (2, 3). In turn, lithium carbonate and other forms of lithium were believed to be as effective in dissolving uric acid in vivo as in vitro, despite a dearth of experiments quantifying the physiologic and toxic effects of lithium. One experiment taken to be "close enough" to simulating in vivo conditions involved soaking a uric acid-studded metacarpal bone in a highly concentrated solution of lithium, the consequences of which were the dissolution of the urate crystals and the recovery of the intact bone. This was taken to be evidence of the potential for lithium to "cure" gout (1, 3).

Tablets of lithium carbonate and chloride were readily available to the public. Many physicians and lay people, however, felt that gastrointestinal symptoms limited pill-form utility (4). In contrast, lithia springs water (mineral water reputed to contain lithium salts) was cheaper, equally available, and less likely to cause gastrointestinal distress. Thus, a valuable new market was opened to producers of mineral waters: persons believing that lithium-containing waters would relieve them of a host of ills.

Testimonials, both real (4, 5) and perhaps forged, were widespread as to both the power and potency of the waters. Claims included the following:

Manadnock Lithia Water (is) recommended for gout, dyspepsia, rheumatism, eczema, sugar diabetes, Bright's disease, gall stones; also reduces temperature in all fevers, and all diseases of the kidneys, asthma, etc. As a beautifier of the complexion, it has no equal (6).

One clinician who tested the effects of Buffalo Lithia Water on a gout-ridden relative found "relief from the threatening symptoms — so prompt and decided as to be almost incredible" (5). But the use of lithia water in gout was rarely the only component of cure:

Encouraged by what he has seen, read, and heard, (the gout-ridden patient) applies himself assiduously to the consumption of potent waters to the exclusion of alcohol, and eats plain food in moderation in place of his usual diet. He drinks lithia at breakfast, and in the forenoon, a bottle or more with his luncheon . . . and finally as a substitute for the usual nightcap. As a result of his abstemiousness . . . but in his own mind of his devotion to large volumes of the powerful solvent, his condition is soon greatly improved (7).

At the turn of the century, analysts examined both the claims and the content of these "natural" spring waters. Using a variety of methods of chemical extraction and spectrophotometric flame emission analysis, most of the lithia waters — though not all — proved to contain little to no detectable lithium (i.e., less than microgram quantities) (1, 7–9). In some cases, lithium compounds were found to have been artificially added. In other springs, fecal contamination was present (6).

These findings led to legal actions that were stimulated by the investigations of the recently created Food and Drug Administration. In one such case, *U.S. vs Buffalo Springs*, the judge noted that to obtain a "therapeutic" dose, consumers may have had to drink from 150,000 to 225,000 gallons of Buffalo Lithia Water per day (6). After the case was brought to the U.S. Supreme Court, the company remained in business, but changed its product name to Buffalo Mineral Springs Water.

Thus, for many years, lithium in water was regarded as no more effective or necessary to human health than any good water. Indeed, lithium acquired a bad reputation after use of lithium chloride as a table salt substitute in the 1940's resulted in what should have been predictable deaths and morbidity (1, 10–13). With the emergence of the pharmacological use of lithium in the therapy of manic-depressive illness (14), there developed a resurgence of interest in naturally occurring lithium.

Drinking Water Lithium and Mental Hospital Admissions

One outgrowth of this renewed interest in lithium was the investigation of the relationship of mental hospital admissions to drinking water lithium content (15–17). These studies may be viewed as addressing the question of whether lithium's therapeutic effects exist at what is generally considered to be homeopathic levels.

The first study, which originated in Texas, concluded that with increasing lithium water content, mental hospital admissions and city homicide rates declined (15–17). Among the assumptions made were that every water lithium level reported by the 27 cities surveyed represented a uniform finding, reproducible at any water-sampling site within that city, and that water-sampling methods were uniform among cities.

Most city water systems, however, are composed of subsystems (18, 19), starting with multiple subsurface and surface reservoirs, and, thus, are subject to differential mineral and soil contaminants. Water is then typically pumped through a primary pumping station where water purification may occur, followed by distribution through major conduits. In turn, there may be smaller substations throughout an area served by the primary pumping station. Alterations in water mineral content may, therefore, occur at several sites. Thus, uniform sampling techniques in large-scale studies may be critical.

A nonuniform method of sample analysis for lithium may also be a source of experimental variability. While most investigators and clinical laboratories use flame emission spectrometry or atomic absorption spectroscopy (20, 21), other investigators have used mass spectrometry with neutron activation (22, 23)—a technique with the ability to reliably detect less than nanogram quantities of lithium. Finally, sample preparation techniques may be a source of variation; for example, porcelain vessels can be a source of lithium contamination (24).

In the Texas study of mental hospital admissions (15, 16), the site, method, and handling of the water lithium levels were not described. The cities were divided into "lithium quintiles," according to reported water lithium levels. These quantities ranged from less than 11 mcg/L to greater than 70 mcg/L. While there was a statistically significant decrease in hospital admissions between cities in the highest quintile versus those of the lowest, there was no apparent relationship between the intermediate quintiles and those of the extremes. One might expect that if there was a true relationship, a smooth or multivariate-governed relationship might exist. No such statistical analysis was undertaken.

Subsequent studies noted this lack of apparent relationship and cogently pointed out technical problems (19). For example, El Paso, classified in the greater than 70 mcg/L quintile, had a variation of lithium content over time and across its distribution system ranging from 18 to 168 mcg/L (19). Furthermore, it was recognized that state hospital admissions varied inversely with distance of the hospital from the cities studied. El Paso (higher lithium water concentration, lower hospital admission rate) was considerably further from the state hospital than Austin (lower lithium water concentration, higher hospital admission rate). It is thus not surprising, in light of these methodological problems, that the sole attempt to replicate this study (17) found no correlation between water lithium content and mental hospital admissions.

Drinking Water Lithium and Cardiovascular Mortality

From the late 1950's to the mid-1960's, a small body of work evolved that noted a geographic and geohydrologic distribution of cardiovascular mortality (25). Studies were performed that attempted to correlate water mineralization—specifically, calcium carbonate content (water "hardness") with cardiovascular mortality (26, 27). For example, in a study that examines the mineral content of ocean waters, one author noted that "states having a seacoast showed higher cardiovascular death rates than states without" (28), a finding for which several etiologies may be found.

In an epidemiologic effort to show that lithium in drinking water protects against atherosclerotic heart disease (AHD), Voors (29) looked for correlations between age-adjusted AHD mortality rates and water hardness and water lithium content in 99 of the 100 largest cities in the United States. The water data were taken from the 1962 Geological Water Supply Survey (30). Negative product-moment correlation coefficients between lithium level in municipal water and AHD mortality were significant at the 5% level in white and nonwhite males and females. On the other hand, the correlation coefficient for water hardness was significant only in white males. Another survey (31), which focused on lithium water content, found a moderate correlation with cardiovascular mortality without examining subpopulations by race or gender. In a third study based in western Texas (32), similar results were obtained. While all of these studies used water lithium as their index of population lithium exposure, none examined the intake of lithium through foodstuffs—a source that is felt to provide up to 35 times more dietary lithium than water (33). Such mass measurements also ignore the presence in the sample of a variable number of persons taking therapeutic doses of lithium (estimated at 25 to 324 mg lithium per day). This subpopulation might, in turn, skew the cardiovascular mortality rates. Therefore, while the consistency of correlations of cardiovascular mortality to the presence of water lithium is interesting, it does not establish a causal relationship. Indeed, a prospective epidemiological study of men in Switzerland found no correlation between lithium water content and cardiovascular mortality (34).

Drinking Water Lithium and Dental Caries

Yet another variation on this theme is found in dental caries research. As is well known, the fluoride content of water plays a distinctive role in caries prevention. In an examination of artesian versus surface water supplies and their effects on the teeth of aboriginal and Caucasian children (35), the authors hypothesized that, in addition to the cariostatic effect of the high fluoride concentration in artesian water, lithium concentration (132 mcg/L) of the artesian water also contributed to a lower caries prevalence among children 6 to 11 years old. However, a similar study using 12 to 14 year olds in western Texas came to an opposite conclusion (36) — that is, when one holds constant the contribution of fluoride to caries prevention, lithium appears to play a cariogenic effect.

Teratogenicity and Oncogenicity of Drinking Water Lithium

Thus, we are left with little evidence for a beneficial effect of drinking water lithium on mental hospital admissions, cardiovascular disease, or dental caries. Are there deleterious effects? For example, are there teratogenic effects at usual drinking water concentrations (usually considered to be well under 1 mg/L)? The answer appears to be no. It should be noted that no studies regarding teratogenesis have been done in areas with very high water lithium concentrations, such as Northern Chile, where lithium levels have been reported to reach 5.2 mg/L (37). Still, the latter is 10 to 20 times lower than a typical daily psychotherapeutic lithium dose in the U.S. (900 to 1200 mg lithium carbonate is approximately equivalent to 169 to 226 mg elemental lithium). While lithium does appear to have teratogenic properties at relatively high exposure levels, there is no support for extrapolating these findings to very low levels. In other words, a lower-limit threshold must exist for the dose-response (teratogenicity) curve. In an analysis of possible toxic effects of an industrial outflow containing 6 to 12 mg/L lithium into drinking water, a lack of evidence is cited that would indicate any teratogenic — or, indeed any toxic — effect on those drinking this water, even at the point of maximal lithium content (38).

A similar lack of clear evidence exists regarding the mutagenicity of typical drinking water lithium levels when utilizing leukocytes as the target cells (39, 40). Results have been both positive and negative in examining in vitro cells of patients treated at therapeutic levels with lithium (41–44). Epidemiological studies of patients treated with lithium reveal no increase in incidence of leukemia or other cancers over baseline incidence (45–47).

The Effect of Low Lithium Diets in Experimental Animals

If there are no clearly established positive or negative effects of naturally occurring lithium, is there any role for lithium in normal life functions? The few existing dietary studies have been the most methodical in addressing this question. In Patt et al. (48) and in the extension of that study (49), 3 generations of rat litters were raised on a low-lithium diet which contained 5 to 10 micrograms of elemental lithium per liter of food. Controls were fed either a "normal" diet constructed by the experimenters (500 micrograms of lithium per liter), or a diet made of standard chow (analyzed as containing 350 micrograms of lithium per liter). While there was no statistically significant decrement in growth patterns of the litter mates, there was a trend toward a decrement in fertility, as demonstrated by an increase in the age of dams at conception in the low-lithium group in the second and third generations. However, the numbers of litters and dams were too small to test the statistical significance of this finding. Similar results have been reported in goats (50). In addition, quail fed low-lithium diets showed decreases in hatchability and shell quality in the offspring of the low-lithium-fed fowl (51).

In addition to examining fertility, the tissue lithium content was examined in the three-generation rat litter studies (48, 49) and the studies of quail (51). In the rat litter studies, endocrine tissues, the corpus striatum, the hippocampus, and ovarian tissues maintained normal levels of lithium on the low-lithium-content diets, while decreases in lithium content below control levels were found in the other examined organs. In the quail study, all tissues were depleted in lithium content, without selective concentration in endocrine or brain tissues. That lithium may be selectively concentrated in certain organs at below-usual physiological levels in the rat suggest that this process does not occur by simple saturation mechanics.

Finally, one study examined the effects of less-than-standard dietary lithium ingestion on lipid metabolism in rats (52). Control rats consuming a stock diet had the highest serum lithium levels. For the remaining animals, lithium was added as a percentage of diet, from none to 0.08% with concomitant increasing serum lithium levels. It is unclear from the article whether the lithium content was a percentage of grams of food, of total intake (i.e., food and water), or some other aspect of the diet. The authors did note a smooth and significant decrement in total serum lipid content with increasing lithium. There were significant decreases in both serum cholesterol and triglycerides as dietary lithium content increased. Thus, at least in the rat, dietary lithium may plan a role in lipid metabolism.

A study which generated questions regarding specific organ uptake of low-level lithium in humans is Benzo et al. (53) observation of an area of endemic goiter in Venezuela. This region has moderate levels of lithium in water (5 to 80 micrograms per liter). The authors noted that while iodine deficiency was well established in the involved communities, the prevalence of goiter varied among them. Since pharmacologic amounts of lithium can have goiterogenic effects (54), they wondered whether variations in drinking water lithium might play a role in the differing prevalences. However, the authors did not quantitate the lithium content of food in this population despite noting it as a potentially significant source of lithium nor did they measure food or water iodine content; consequently, no firm conclusions could be reached. Given this report and the rat-litter study results, further studies would be of interest to evaluate a possible effect of subtherapeutic amounts of lithium on thyroid function.

Summary and Conclusions

This article has summarized findings in disparate areas of medical inquiry whose common theme is the effects of nonpharmacologic lithium intake. Many of the studies are plagued with methodological problems. Studies that are sound have yet to be replicated and extended. And, while the studies theorize about low-level lithium effects, none have addressed the mechanism of action. One may specu-late, for example, that the absence or presence of lithium may be affecting body systems through interactions with sodium, potassium, and magnesium, or through interactions with cyclic AMP (42). Alternatively, while naturally occurring lithium may be selectively taken up by certain rat tissues or may have a role in fertility, it may also be that lithium is merely present and measur-able, but not playing a causative role. More research on these topics is needed to determine whether lithium, like some other minerals, is an essential element in human function.

References

1. Good, C.A. (1903). *On the action of lithium*. Chicago: Lakeside Press.
2. Garrod Sir, A.B. (1859). *The nature and treatment of gout and rheumatic gout*. London: Walton and Maberly.
3. Haig, A. (1900). Uric acid as a factor in the causation of disease (5th ed.). Philadelphia: P. Blakiston.
4. Morse, W.H. (1887). A contribution to the study of the therapy of lithia water. *The Medical Age, 5*, 438–439.
5. Williamson, J.W. (1878-79). Buffalo lithia waters for uremia. *Virginia Medical Monthly, 5*, 898–899.
6. Cramp, A.J. (1921). *Mineral waters. Nostrums and quackery*. Chicago: American Medical Association.
7. Harrington, C. (1896). On the action of commercial lithia waters. *Boston Medical and Surgical Journal, 135*, 644–645.
8. Leffmann, H. (1910). Lithia waters as therapeutic agents. *Monthly Encyclopedia and Medical Bulletin, 111*, 138–144.
9. Waller, E. (1890). Determination of lithia in mineral waters. *Journal of the American Chemical Society, 12*, 214–223.
10. Cleaveland, S.A. (1913). A case of poisoning by lithium, presenting some new features. *JAMA, 60*, 722.
11. Corcoran, A.C., Taylor, R.D., & Irvine, H. (1949). Lithium poisoning from the use of salt substitutes. *JAMA, 139*, 685–688.
12. Hanlon, L.W., Romain, M., Frank, J., et al. (1949). Lithium chloride as a substitute for sodium chloride in the diet. *JAMA, 139*, 688–690.
13. Aaron, H. (1949). Dangerous drugs. *Consumer Reports, 14*, 171–173.
14. Cade, J.F. (1949). Lithium salts in the treatment of psychotic excitement. *Medical Journal of Australia, 36*, 349–352.
15. Dawson, E.B., Moore, T.D., & McGanity, W.J. (1970). The mathematical relation-ship of drinking water lithium and rainfall to mental hospital admission. *Diseases of the Nervous System, 31*, 811–820.
16. Dawson, E.B., Moore, T.D., & McGanity, W.J. (1972). Relationship of lithium metabolism to mental hospital admission and homicide. *Diseases of the Nervous System, 33*, 546–556.
17. Voors, A.W. (1972). Drinking-water lithium and mental hospital admissions in North Carolina. *North Carolina Medical Journal, 33*, 597–602.
18. Safe Drinking Water Committee. (1972). Chapter II: Elements of public water sup-plies. *Drinking Water and Health*, Vol. 4 (pp. 9–17) Washington, DC: National Academy Press.

19. Trieff, N.M., Severn, M.F., Rao, M.S., et al. (1973). Analysis for lithium in Texas drinking waters. *Texas Reports on Biology and Medicine, 31*, 55–78.
20. Christian, G.D. (1975). Biological fluids. In J.A. Dean, T.O. Rain (Eds.), *Flame Emission and Atomic Absorption Spectrometry, 3* , 367–413.
21. Christian, G.D., & Feldman, F.J. (1975). The alkali metals. In G.D. Christian, & F.J. Feldman (Eds.), *Atomic absorption spectroscopy: Applications in agriculture, biology and medicine* (pp 215–234). New York: Wiley and Sons.
22. Clark, W.B., Koekebakker, M., Barr, R.D., et al. (1987). Analysis of ultra-trace lithium and boron by neutron activation and mass spectrometer measurement of 3-He and 4-He. *Applications of Radiation Isotopes, 38*, 735–743.
23. Clark, W.B., Webber, C.E., Koekebakker, M., et al. (1987). Lithium and boron in human blood. *J. Lab. Clin. Med., 109*, 155–158.
24. Versieck, J., Vanballenburghe, L., DeKesel, A., et al. (1987). Accuracy of biological trace-element determinations. *Biological Trace Elements Research, 12*, 45–54.
25. Kobayashi, J. (1957). Geographical relationship between the chemical nature of river water and death from apoplexy. *Berichte du Ohara Institut fur Landvirtsch Biologie, 11*, 12–21.
26. Schroaeder, H.A. (1960). Relation between mortality from cardiovascular disease and treated water supplies. *JAMA, 172*, 1902–1908.
27. Morris, J., Crawford, M.D., & Heady, J.A. (1961). Hardness of local water supplies and mortality from cardiovascular disease. *Lancet, 1*, 860.
28. Schroaeder, H.A. (1966). Municipal drinking water and cardiovascular death rates. *JAMA, 195*, 125–129.
29. Voors, A.W. (1970). Lithium in the drinking water and atherosclerotic heart death: Epidemiological argument for protective effect. *American Journal of Epidemiology, 92*, 164–171.
30. Durfor, C.N., Becker, E. (1964). Public water supplies of the 100 largest cities in the U.S. 1962 Geological Water Supply Paper 1812, Washington, DC.
31. Blachy, P.A. (1969). Lithium content of drinking water and ischemic heart disease. *New England Journal of Medicine, 281*, 682.
32. Dawson, E.B., Frey, M.J., Moore, T.D., et al. (1978). Relationship of metal metabolism to vascular disease mortality rates in Texas. *American Journal of Clinical Nutrition, 31*, 1188–1197.
33. Livingston, H.D. (1970). Lithium depletion and atherosclerotic heart disease. *Lancet, 2*, 99.
34. Kromhout, D., Wibowo, A.A.E., Herber, R.F.M., et al. (1985). Trace metals and coronary heart disease risk. *American Journal of Epidemiology, 122*, 378–385.
35. Schamschula, R.G., Cooper, M.H., Agus, H.M., et al. (1981). Oral health of Australian children using surface and artesian water. *Community Dental and Oral Epidem., 9*, 27–31.
36. Curzon, M.E.J., Richardson, D.S., & Featherstone, J.D.B. (1986). Dental caries prevalence in Texas school children using water supplies with high and low lithium and fluoride. *J. Dent. Res., 65*, 421–423.
37. Schull, W.J. (1980). Food, water and genes. *Fukuoka Igaku Lasshi, 71*, 47–60.
38. Jefferson, J.W. Potential effects of lithium in water at levels discharged by Foote Mineral Company at Frazer, Pennsylvania Plant, Appendix D. Re: NPDES Draft Permit No. PA0014222 (unpublished).
39. Budd, J.L., & Rossof, A.H. (1980). Drinking water lithium levels fail to predict for the incidences of acute or chronic granulocytic leukemia. *Adv. Exp. Med. Biol., 127*, 410–416.

40. Aglietta, M.Z., Piacibello, L., & Piacibello, W. (1981). Effect of lithium on normal and chronic granulocytic leukemia colony forming cells. *Experimentia, 37,* 1340-1341.
41. Sabine, H., Swierenga, H., Gilman, J.P.W., et al. (1987). Cancer risk from inorganics. *Cancer and Metastasis Reviews, 6,* 113-154.
42. Bille, P.E., Jensen, M.K., Jensen, J.P.K., et al. (1975). Studies on the hematologic and cytogenetic effect of lithium. *Acta Med. Scand., 198,* 281-286.
43. Dey, S.K. (1985). Effects of lithium carbonate on the bone marrow chromosomes of the rat. *J. Env. Biol., 6,* 103-106.
44. Matsushima, Y., Hazama, H., & Kishimoto, K. (1986). Chromosome examination of patients under lithium therapy. *Japanese Journal of Psychiatry and Neurology, 40,* 625-630.
46. Lyskowski, J., & Nasrahla, H.A. (1981). Lithium therapy and the risk for leukemia. *Br. J. Psych., 139,* 2561.
47. Resek, G., & Oliveri, S. (1983). No association between lithium therapy and leukemia. *Lancet, 1,* 940.
48. Patt, E.L., Pickett, E.E., O'Dell, B.L. (1978). Effects of dietary lithium levels on tissue lithium concentrations, growth rates and reproduction in the rat. *Bioinorganic Chemistry, 9,* 299-310.
49. Burt, J., Dowd, R.P., Pickett, E.E., et al. (1982). Effect of low dietary lithium on tissue lithium content in rats. *Fed. Proc.* (abstracts), *41,* 460.
50. Nielsen, F.H. (1984). Ultrarace elements in nutrition. *Ann. Rev. Nutr., 4,* 21-41.
51. Hempe, J.M., Savage, J.E., & Pickett, E.E. (1985). Effect of low dietary lithium on growth, reproduction and concentration of lithium of Japanese quail. *Poultry Science, 64,* 116.
52. Fleishman, A.I., Lenz, P.H., Bierenbaum, M.L. (1974). Effect of lithium upon lipid metabolism in rats. *Journal of Nutrition, 104,* 1242-1245.
53. Benzo, Z., Fraile, R., Cevallos, J.L., et al. (1984). Urinary lithium excretion in endemic and nonendemic goiter areas in Venezuela (Preliminary Report). In P. Bratter, & P. Schramel (Eds.), *Trace element analytical chemistry in medicine and biology,* Vol. 3 (pp. 467-474). Berlin: Walter De Gruyter.
54. Hyman, S.E., Arana, G.W. (1987). *Handbook of psychiatric drug therapy.* Boston: Little, Brown Company.

3
Lithium in Plants

Charles E. Anderson

Introduction

The discovery of lithium (Li) in 1817 is attributed to J.A. Arfwedson. Bervzelius proposed the name for the element based on the Greek word *lithos*, meaning stone (2). Lithium is widely distributed in relatively small quantities throughout the earth's crust. A number of studies have involved the determination of total Li in soils. Steinkoenig (3) sampled soils in the United States and found 10 to 100 ppm Li. Subsequently, Soviet investigators reported lower levels ranging from 10 to 50 ppm Li (4, 5, 6), while Swaine (7) found wide variation in mineral-rich soils with 8 to 400 ppm Li. However, it is improbable that the total Li in soils is available for uptake by plants. The Li that can be extracted from soils in California ranged from 0.1 to 0.9 ppm (8), while higher levels, 0.4 to 2.5 ppm, were contained in saline Indian soils (9). Insufficient information is available concerning the relationship between Li which can be extracted from the soil, and the quantity of Li absorbed by plants. Aldrich et al. (10) found increases of Li concentrations in lemon leaves with decreasing soil pH in greenhouse studies.

As a result of lithium-containing mineral degradation, Li^+ ions are leached from the soil by rainfall and enter the water supply. The mean Li concentration of seawater is 0.2 ppm, yet some natural brines range to 400 ppm (11). In general, fresh water contains smaller quantities, averaging 0.03 ppm Li, but a number of investigations revealed higher levels associated with certain irrigation waters (12, 13, 14).

The occurrence of Li in plant tissue was described first in 1872 (12) and this was followed by a number of other reports (16, 17, 18). Bertrand (19–23) analyzed hundreds of species, and in general found concentrations of 1 ppm Li in plant tissue. The Magnoliopsida averaged slightly higher than the Liliopsida. Subsequent cation studies (24, 25) provided information concerning the uptake of Li by a number of cultivated plants. It became apparent that certain families tended to accumulate higher Li levels in their tissues. Ezdakova (26) observed that members of the Solanaceae and Ranunculaceae established high tissue concentrations of Li. In fact, *Lycium* accumulated substantial quantities of Li, even when Li was not abundant in the soil. This was substantiated in recent reports in the United States (27, 28, 29). As a result of this differential accumulation certain species have been used, particularly in the Soviet Union, in geobotanical prospecting for Li ore deposits (30).

The versatility of Li and its compounds is tremendous, and a full range of their industrial applications has hardly become apparent. Major industrial uses at this time include light-weight alloys, lubricating greases, air-conditioning and dehumidification systems, polymerization catalysts, detergent compounds, glass and ceramics fluxing, medicinal uses in the treatment of manic-depressive psychoses, and many others. This trend towards increased utilization may lead to the release of Li into the environment and the possibility of air, soil, and water contamination. However, at present, the only documented cases of naturally occurring Li toxicity have been in citrus groves of the southwest United States using irrigation waters (31, 32). This has occurred due to continuous irrigation leading to concentration of Li in soils through evapotranspiration. No industrial source has been implicated with these particular occurrences. Due to the release of Li into water systems as waste from industrial processes (33), and natural occurrences of toxic levels of Li in irrigation waters, Bradford (12) suggested the need for routine analysis of water supplies for Li by agencies responsible for maintaining water quality standards. However, a recent report found virtually no danger that toxic amounts of Li could be ingested by humans from environmental sources (33).

Phytotoxicity

Lithium phytotoxicity was first observed in 1871 during experiments involving the role of K in plant nutrition. Solutions containing Li produced death of buckwheat and biomass reductions in rye (34). Gaunersdorfer (35) proposed that Li is transported to leaves by transpiration, where it accumulates, and causes injury. Voelcker (36–42) described the occurrence of Li toxicity during trials of trace metals for use in agriculture. Reductions in biomass accumulation were produced by 19 kg Li ha^{-1} in wheat (*Triticum aestivum*) and barley (*Hordeum vulgare*). Germination was prevented by 77 kg ha^{-1} in clover (*Trifolium pratense*) and 19 kg ha^{-1} in pea and mustard. Unfortunately, the method of application of Li to the soil surface in solution left much to be desired. This and similar shortcomings in the design of early Li toxicity experiments left many results unclear and interpretations precarious at best. Later, more methodic experiments by Voelcker (41) established a toxic threshold for Li in wheat at 30 ppm in soil.

Contemporary Italian investigators were again testing the possible replacement of K with Li in nutrient solutions (43, 44). Pronounced toxic effects were encountered in bean (*Phaseolus* sp.) and oat (*Avena sativa*) at 189 ppm Li, while tobacco (*Nicotiana tobacum*), although reduced 33% in biomass, was described as having some tolerance to Li. Tobacco (*T. tobacum*) accumulated Li in tissues at high concentrations. Lithium at 252 ppm produced injury, in decreasing order, in tomato (*Lycopersicon esculentum*), mustard (*Brassica hirta*), hemp (*Cannabis sativa*), sunflower (*Helianthus annuus*), flax (*Linum usitatissimum*), vetch (*Vicia sativa*), and corn (*Zea mays*). In particular, corn (*Z. mays*) showed no reductions in biomass when subjected to this constant level of Li. Experiments in subsequent years described Li toxicity in olive (*Olea* sp.) (45), wheat (*T. aestivum*) (46), buckwheat (*Fagopyum esculentum*) and barley (*Hordeum vulgare*) (47), and cabbage (*Brassica olearacea*), celery (*B. pekinensis*), and lettuce (*Lactuca sativa*) (48).

Frerking (49), after observing the toxic effects of Li, concluded that Li prevented the uptake of Calcium (Ca) by plants. However, Haas (50) found adequate concentrations of Ca in *Citrus* sp. treated with Li. He, none the less, proposed again that Li perturbed Ca requirements in plants, but that the antagonism occurred within the plant. In his study, injury was produced in sweet orange at 3 ppm Li in soil, while in solution culture lemon (*Citrus limon*) and sweet orange (*C. sinensis*) leaves displayed necrosis at 1 and 2 ppm Li, respectively. Subsequently, other experiments (1, 51) demonstrated that Li injury could be mollified by Ca supplements.

Experiments in Germany (52) concluded Li could not satisfactorily replace K, and caused reductions in fresh weight biomass at 69 ppm Li in solution with oats (*Avena sativa*), wheat (*T. aestivum*), rye (*Secale cereale*), and barley (*H. vulgare*). Similar reductions occurred at 6.9 ppm in pea (*Pisum sativum*), buckwheat (*Fagopyun esculentum*), mustard (*B. hirta*), red clover (*T. pratense*), and alfalfa (*Medicago sativa*). Furthermore, Sharrer (53) and later investigators (54) concluded Li was the most toxic of the alkali metals.

The initial report (31) in 1951 of naturally occurring Li toxicity described injury to citrus in Santa Barbara County, California. This became prevalent upon increased irrigation of grooves from ground wells containing 45 to 80 ppb Li. Studies (75, 76) followed to determine the possibility that these concentrations would affect other agricultural crops. Results showed the concentrations necessary to produce injury were well above those in the contaminated soil and water. Responses to and accumulation of lithium in tissues were found to be species-dependent, and the plants were classified according to tolerance. Avocado (*Persea americana*), soybean (*Glycine max*), and sour orange (*Citrus* sp.) were most sensitive, while sweet corn (*Z. mays*) and rhodesgrass (*Chloris gayana*) exhibited resistance. The investigators noted a parallel between lithium tolerance and sodium tolerance in plants. In general, tolerant plants accumulated much higher levels of Li in their tissues, except sweet corn (*Z. mays*), which contained only small quantities. Analysis of tissue revealed no differences in the essential nutrient composition of the plants despite the extreme variation in Li accumulation.

Concentrations of Li in tissues associated with Li toxicity for a number of species and treatments are presented in Table 3.1. This tabulation, although not intended to be a complete record of all Li toxicity experiments, includes citations for those experiments which presented data on dosage, tissue concentrations, and biological response. The toxicity of Li depends on both the rate of accumulation and the affinity for reaction with tissue components as determined by the species. A recent study (57) of the effect of Li on seed germination showed reductions in percent germination at 5 ppm Li with rice (*O. sativa*) and gram (*Cicer arietinum*), while no decrease occurred in barley (*H. vulgare*) or wheat

TABLE 3.1. Species, tissue sampled, concentration of lithium in tissue (ppm, dry weight), dose, and biomass reduction (% g dry weight) of plants exhibiting lithium toxicity.

Species	Tissue sample	Li ppm	Dose	%-Biomass reduction	Ref.
avocado	leaf	120	3 ppm substrate	25	56
Persea americana	root				
	petiole				
barley	shoot	1131	500 ppm substrate	66	70
Hordeum vulgare					68
bean					
Phaseolus vulgaris					
dwarf red kidney	leaf	140	12 ppm substrate	25	56
	root	520			
	petiole	140			
	shoot	72	5 ppm solution	14	56
bush	leaf	546	25 ppm substrate	15	67
Phaseolus sp.	stem	336			
	leaf	226	25 ppm substrate	32	68
	stem	519			
	leaf	19	0.5 mM solution	16	68
	stem	48			71
	root	24			12
beet, red	leaf	4500	35 ppm substrate	25	56
Beta sp.					
sugar	shoot	300	1 mM solution	18	72
bluegrass	root	2	5 ppm substrate	–	73
Poa ampla	shoot	11			72
celery	leaf	900	–	30	61
Apium graveolens					
corn, sweet	leaf	160	70 ppm substrate	25	56
Zea mays					
cotton	leaf	1100	25 ppm substrate	25	56
Gossypium sp.	root	250			
G. barbadense	leaf	750	50 ppm substrate	5	74
('giza 45')	stem	224			23
('giza 68')	leaf	6.7	2.0 μM solution	19	75
	stem	3.2			96
G. hirsutum	leaf	1947	100 ppm substrate	74	74
('Acala SJ-2')	stem	1627			93
dallis grass	leaf	340	25 ppm substrate	25	56
Paspalum diletatum					
grape	leaf	300	12 ppm substrate	25	56
Vitis sp.	root	400			
grapefruit	leaf	14	irrigation water	a[1]	31
Citrus × *paradisi*					
'Marsh'	leaf	60	irrigation water	a[1]	32
lemon	leaf	12	irrigation water	a[1]	31
Citrus limon	rind	2	irrigation water		
	juice	0.5			
lily, Easter	leaf	157	irrigation water	a[1]	77
Lilium longiflorum					

TABLE 3.1. (*Continued*)

Species	Tissue sample	Li ppm	Dose	%-Biomass reduction	Ref.
orange, navel	leaf	23	irrigation water	a[1]	115
Citrus sp.					
sour	leaf	180	8 ppm substrate	25	56
C. aurantium	root	50	4 ppm solution	36	56
	shoot	256	7 ppm substrate	25	56
sweet	leaf[2]	25	2 ppm substrate	a[1]	31
C. simensis	leaf[3]	220			
	stem[4]	25			
	stem[5]	15			
	root[4]	20			
	root[5]	5			
'Valencia'	leaf	14	irrigation water	a[1]	31
Citrus sp.					
soybean	leaf	40	7 ppm substrate	25	56
Glycine max	root	80			
	petiole	40			
tomato	root	100	12 ppm substrate	25	56
Lycopersicon lycopersicum	petiole	380			
	shoot	316	5 ppm substrate	22	56
'Tropic'	leaf	80	2 ppm solution	a[1]	78
Lycopersicon esculentum					
wheatgrass,	root	27	15 ppm substrate	38	73
crested	shoot	150		22	
Andropyron desertorun					
'Whitmar'	root	4	2.5 ppm substrate	60	73
Andropyron sp.	shoot	6		54	

[1]Toxic symptoms exhibited on leaves; [2]young; [3]mature; [4]bark; [5]wood.

(*T. aestivum*), even at 100 ppm. Root elongation, following germination, was severely retarded in gram at 5 ppm Li and in the other species at 10 ppm.

Fungal infections were reduced after injections of Li into infected trees (58, 59). Later, the toxic effect of Li to fungi was established during experiments with *Aspergillus niger* (60). Further experiments investigated the use of Li as a fungicide and bactericide (61–64). In general, application rates controlling the infection were also phytotoxic and when rates were reduced to nonphytotoxic levels, the disease prevailed. A recent investigation (65) found 5 uM Li inhibiting appressorial and haustoria development in *Erysiphe graminis* f.s. *hordei* on barley (*H. vulgare*). They were unable to determine whether Li affected the pathogen, host, or both. Tobacco (*Nicotiana tobacum*) plants were grown in 0, 10, 20, 30, and 40 ppm Li wt/wt in soil in a greenhouse. Plant leaves developed Li toxicity symptoms in the 40 ppm treatment after two weeks. At that time plants were inoculated with one of a series of RNA and DNA viruses. Viral titers were assessed on all plants after one week. In no instance did lithium-treated plants reduce viral multiplication in tobacco leaf tissues (69).

Wallace et al. (66) determined that the frequency distribution of Li accumulation in corn (*Z. mays*) is log$_e$ normally distributed. The study was conducted in soil combinations with a number of other trace metals. In other work (67), synergism was not found in the toxic effects produced by Li and Cd, and Li, Cd, Cu, and Ni in bean (*P. vulgaris*). In both cases, toxic effects on biomass production were less than additive and Li accumulation in leaves decreased when in combination with the other elements, while stem concentrations increased. Finally, the use of the chelating agent, ethylenediaminetetraacetate (EDTA), resulted in slightly increased Li concentrations in bean (*P. vulgaris*) cultured in nutrient solution (68).

Field observations in oak-hickory forests around sources of Li emissions have lead to several conclusions regarding the sensitivity of native vegetation to Li from both air and soil. First, sensitive plants are affected very early in the growing season. Thus, the physical response is often one of leaf cupping, and marginal necrosis on young leaves. Species such as chestnut oak (*Quercus prinus*), Mockernut hickory (*Carya tomentosa*), and black cherry (*Prunus serotina*) are species which indicate Li presence in this forest setting (see Table 3.3). As the season progresses Li concentrations continue to build in plant leaf tissue. At that point more species are injured, but the injury pattern produces less leaf curling, since the affected leaves are already fully expanded. In this case the symptom is usually marginal chlorosis followed by necrosis. In some species such as blueberry (*Vaccinium* sp.) or river cane (*Arundinaria tecta*) the injury begins at the leaf tip and progresses basipetally. Finally, later in the growing season, tolerant species such as blackjack oak (*Quercus marilandica*), willow oak (*Q. phellos*), mulberry (*Morus rubra*), sweet gum (*Liquidambar styraciflua*), and Virginia pine (*Pinus virginica*) develop necrosis at the leaf tips. Monocots such as Johnson grass (*Sorgum halepense*), bromesedge (*Andropogon virginicus*), and some of the panicum grasses develop tip necrosis at this time. The most tolerant plant species in the oak-hickory association are sweet gum (*L. styraciflua*) and Virginia pine (*P. virginica*) (see Table 3.3).

Growth Stimulation

Although the previous data clearly demonstrate the detrimental effects of Li to plants, a body of evidence exists which shows that Li stimulates the growth of plants at lower concentrations.

In 1904, Nakamura (79) reported lithium-induced increases in biomass with pea and barley. The possibility that Li has beneficial characteristics was first suggested by Voelcker (37). In the study wheat seeds soaked in a Li solution show biomass increases. After establishing a toxicity threshold for Li in wheat, Voelcker (41) noticed Li exerted a stimulating influence at subtoxic concentrations. Later Brenchley (47) showed an increase in barley (*H. vulgare*) biomass at 0.02 ppm Li in nutrient solutions. Scharrer and Schropp (52) produced increases in fresh weights in rye (*S. cereale*), barley (*H. vulgare*), and oats (*A. sativa*), when those species were grown at concentrations less than 1 mM Li.

Though not fully investigated, Hanace (80) perceived enhanced sugarcane growth in Hawaiian soils containing quantities of Li. During field studies, Puccini (81) measured increases in the vegetative development of carnation after treatment with Li. Rice (*O. sativa*) seedlings in solutions containing 5 ppm Li displayed enhanced growth (57). Likewise, cotton (*Gossypium barbadense* and *G. hirsutum*) exhibited biomass increases in a number of experiments (74, 75, 68) and was found to accumulate high tissue concentrations of Li.

Beet (*Beta vulgaris*) was shown to accumulate large quantities of Li (56). Ell-Sheikh et al. (72) suggested Li might enhance growth at concentrations less than 2 ppm in nutrient solution. Furthermore, two cool season grasses showed large increases in yield at 2.5 ppm Li in soil (73).

A number of Soviet investigators have asserted a beneficial role of Li in plant growth and development. Based on the positive influence of Li on chlorophyll content, photosynthesis, carbohydrate and nitrogen metabolism, respiration, nucleic and organic acid content, and biomass accumulation, Okhrimenko and Kuz'menko (82) concluded that Li is beneficial to the growth of certain plants. In fact, work was initiated on the formulation of lithium-containing fertilizers for use in agriculture. Results (83) of these fertilizer trials with potato found intensification of sugar synthesis in leaves and its conversion to starch in tubers. This lead to increased starch levels and biomass in the tubers.

Tissue concentrations of Li in plants exhibiting lithium-induced growth stimulation for a number of species and treatments are presented in Table 3.2. Again, this is not a catalog of all experiments reporting lithium-induced growth, and contains only those reports in which comparisons can be made for dosage, tissue concentrations, and biological response.

In reviewing the data on lithium-induced growth stimulation, it has become clear that all these increases are not striking and in many cases may have been obscured by the inherent variation between plants. This has led a number of investigators to describe increases as statistically insignificant. However, these results may very well be economically significant in agricultural applications. There seems to be sufficient cause for an economic evaluation of Li as an agrochemical.

Physiological and Biochemical Effects

Lithium is not considered an essential element in biotanical systems. That is, it is not required for a plant to complete its life cycle. However, while Li may enter a plant and cause no noticeable effect, it may interfere with general plant processes, directly attack specific metabolic pathways, or promote growth. The effects seem to be a product of Li concentration coupled with the genetics of a specific plant. Different varieties of beans respond very differently to identical Li concentrations and growth conditions (1). McStay (84) produced curves for

TABLE 3.2. Species, tissue sampled, concentration of lithium in tissue (ppm, dry weight), dose, and biomass increase (% g dry weight) of plants exhibiting lithium-induced growth stimulation.

Species	Tissue sampled	Li ppm	Dose	%-Biomass increase	Ref.
barley	root	–	0.02 ppm solution	233	67
Hordeum sp.	shoot			135	
	plant		2 ppm substrate	18	79
bluegrass,	root	2	2.5 ppm substrate	23	73
Sherman	shoot	1			
Poa sp.					
cotton	leaf	369.5	25 ppm substrate	8	94
'Giza 45'	stem	98.7		5	
Gossipium bardadense					
'Giza 68'	leaf	4.1	2 µM solution	12	95
G. bardadense	stem	1.7		16	
	root	2.0		25	
'Acala SJ-2'	leaf	373.5	25 ppm substrate	13	94
G. hirsutum	stem	50		21	
'Acala 442'					
G. hirsutum	leaf	587	25 ppm substrate	15	94
pea	plant		2 ppm substrate	9	79
Pisum sp.					
rice	root		5 ppm substrate	15[1]	57
Oryza sativa	shoot			6[1]	
tomato	leaf[2]	291.6	2 ppm substrate		82
Lycopoersicon lycopersicum	leaf[3]	703			
	leaf[4]	765.8			
	stem	20.8			
	root	32.7			
	flower	55.8			
	fruit	5.1		16.7	
	seed	9.3			
wheat	plant		10 ppm substrate	28	41
Triticum sp.	plant		20 ppm substrate	14	
wheatgrass,	root	2	2.5 ppm substrate	66	93
Sherman	shoot	4		23	
Agropyron sp.					

[1]percent increase in length; [2]young; [3]middle; [4]mature.

several plants which showed two distinct detrimental peaks, as well as growth stimulation. Anderson (69) has produced similar results in tobacco (*N. tobacum*) in field trials and tissue culture. One logical conclusion of this type of data is that Li affects more than one plant system. For example, there may be an effect on the membrane transporting system controlling entry of ions into a plant cell. After entering the cytoplasm Li may directly attack various metabolic systems. All this seems to depend on Li concentration in plant tissue, and the time in which the Li impacts on the tissues in question. Growth stimulations may result from stimulated uptake of other ions which participate in various metabolic reactions.

Generally Li is readily absorbed by roots and translocated to stems and leaves. Once in leaves it is generally immobile (85). Some work (86) has shown a low rate of translocation of Li in the phloem. Hinz and Fischer (87) found Li similar to the alkaline earth metals, Ca, Sr, and Ba, in its very slight mobility in the phloem. Their work suggested that the regulation of movement occurred at the sites of phloem loading. One would conclude from this work that under some conditions Li may be transported out of leaves.

Jacobson et al. (88) found Li^+ to be absorbed less than other alkali metal cations in excised barley (*H. vulgare*) and pea (*P. sativum*) roots. Furthermore, Li decreased the absorption of K, but the presence of Ca almost entirely inhibited the absorption of Li and returned K uptake to normal levels. They assumed that Li interfered with K uptake at an absorption reaction site, and that the action of Ca was to prevent access of Li to this site. However, Epstein (51) concluded that Li and Ca shared binding sites after he showed that the uptake of these two ions was competitive. He stated that the sites transporting Li are distinct from those which transport K. Calcium tended to increase K absorption in the presence of Li, while decreasing Li absorption. Epstein found it strange to assume two such dissimilar cations as Ca^{++} and Li^+ to have common binding sites. However, Li^+ ions resemble the alkaline earth cations more than the other alkali cations, with respect to some of the properties that determine chemical affinities, including affinity for carrier sites involved in active ion transport. Amendments of Ca, as $CaCO_3$ were partially effective in reducing Li toxicity in bean (*Phaseolus vulgaris*). The reduction was in leaf area injured (1).

Lithium perturbs a number of physiological processes. It inhibited the characteristic rise in respiration associated with "wound-response" in root slices of chicory (*Cichorium intybus*) (89). Floral induction in duckweed (*Lemma gibba* and *L. perpusilla*) is reduced in the presence of Li, suggesting an influence on phytochrome activity (90). Circadian rhythm of petal movement in *Kalanachoe blossfeldiana* is disrupted by Li (91, 92). Lithium alters thigmomorphogenetic response in beggarticks (*Bidens pilosur* var. *radiatus*) (93, 94, 95) and *Bryonia dioica* (96). In *Byronia dioica* the inhibition was correlated with the suppression of a specific-cathodic isoperoxidase, characteristic of the response. Lithium also decreased the degree and speed of opening and closing of stomata in geranium (*Pelargonium* x *hortorum*) (64). McStay (1) found Li increased stomatal diffusion resistance in bean (*P. vulgaris*) at certain concentrations. These increases were temporary, but recovery was greater at lower Li substrate concentrations. Seismonastic closure of leaflets of sensitive plant (*Mimosa pudica*) was inhibited by Li (1). Lithium inhibited the induction of the "aging" process in leaf discs of geranium (*Pelargonium zonale*), which is usually measured by the development of a methylglucose transport system. This system is not abundant in cells *in situ*, but is synthesized in leaf discs (98).

Lithium affects chlorophyll content (82, 99, 100), photosynthesis (71, 101, 82, 99, 100), respiration (101, 82), and various other metabolic pathways (82, 102, 103). The expression of these effects may be observed in leaf chlorosis of any of the general growth phenomenon previously sited. Lithium may interact with the following cell targets: cell carriers, receptors, and molecules which normally bind other cations; proteins (enzymic, structural, and regulative) which depend on local concentrations of other cations and change conformational state in the presence of Li; and cell membranes, affecting the size of pores and groups that bind ions (104).

To understand this array of responses several explanations have been set forth. In biological systems Li is stable only in the ionic form. Due to a relatively small ionic radius, Li^+ possesses a high electron charge density, making it strongly polarizing, which produces a relatively large, stable, hydrated ion. Vlasyuk et al. (105) proposed that Li affected the specific conformations of macromolecules by influencing their structural bonds with water. They proposed that low Li concentrations increased the amount of water adsorbed to macromolecules, but at abnormally high concentrations Li competed for water and acted as a dehydrating agent. Evidence supporting this hypothesis may come from the work of Hassan (106), who stated that the large degree of hydration for Li^+ ions related to changes in cytoplasmic structure associated with increases in the size of radish root cortical cells after treatment with Li. Also, Li was found to specifically perturb the stacking equilibria in dinucleotide phosphates and polynucleotide. It was concluded that these effects resulted from a change in the hydration shells of the nucleic acids in the presence of Li (107).

Nucleic acids possess many phosphate groups which have a strong ability to bind cations. It is known that the condition of DNA condensation depends on the Na/K ratio and in this respect the Li^+ ion can bind more readily, and lead to changes in the conformation of DNA (2). A review by Vlasyuk, et al. (105) suggests an influence of Li on nucleic acids and protein metabolism. A direct effect of Li on DNA was shown by Okhrimenko et al. (82). Effects on transcription into RNA (108), and translation into protein (109, 110, 111) have also been shown. In general, Li is said by these authors to have its primary effect in influencing the stability and conformational state of DNA, and this in turn affects the metabolism of RNA and proteins. However, as we have already seen, Li may have both a specific effect due to its physicochemical properties, and a nonspecific effect resulting from its similarity to other monovalent cations. This suggests how the effect of Li depends on the dose, and may cause both positive and negative responses.

Tubulin polymerization is promoted by Li, which protects microtubules from the depolymerizing effect of colchicine and vinblastin (112). However, the nuclei of root apical meristem cells of corn (Z. mays) produce abnormal division figures when grown in weak Li solutions (76). At higher concentrations, distortion of metaphase chromosome arrangement, prevention of cell division, and pronounced clumping of chromatin occurs. These studies present apparent opposing views about the effects of Li on DNA and cell replication. However, the adverse responses may be secondary manifestations of some primary effect elsewhere in the system. In both greenhouse and field observations cell replication is influenced. The cells in the margins of leaves of plants like chestnut oak (Q. prinus) are inhibited in both cell division and cell expansion during early leaf growth

(69). The leaves of sweet gum (*L. styraciflua*) are smaller, and abnormally shaped under severe Li stress. This loss of morphogenetic control suggests regulators of cell polarity such as microtubules and Ca^{++} are interfered with in the presence of Li (69). Most certainly, cell replication is influenced in some plants when Li is present in appropriate quantities.

Johnson (113) stated that Li does not strongly bind to proteins. Due to rapid Li exchange, protein structure would not be seriously altered, even when bound at high concentrations. However, there are reports showing the specific bonding ability of Li^+ ions to amides (114), peptides (115), and lipids (116). The fact that Li forms these complexes may influence the ability of proteins to react.

Observations by Stracher (117) illustrating the effects of Li on the physical properties of an enzyme suggested that the association of relatively large hydrated cations with enzymes caused a loosening of conformation and resulted in exposure of reactive groups. A hypothesis by Evans and Sorger (118) proposed a mechanism for the action of univalent cations on enzymes, in which the conformation of univalent cations-activated enzymes is flexible and depends upon the cationic environment. The majority of univalent cation-activated enzymes function in environments containing K. When these enzymes are in an environment where the relatively large cation Li^+ predominates, conformations may be assumed that are not appropriate for catalysis. It was also deemed logical that the conformation of certain enzymes might be less specific, and less critical, and in these cases activation would be possible by Li. It was noted that although hydrated ionic radii seem to contribute most greatly to enzyme specificity, there is a need to consider the chemical properties of the ions involved.

An interpretive model was proposed to explain a number of these cited occurrences (98, 119, 120). According to a "relay and amplification" mechanism, some signals in a physiological response modify cellular pumping or leaking of ions such as hydrogen or sodium. This in turn modifies the K level in cells. The biosynthesis of a few ion-specific proteins is possible only above a critical K level in cells. Accumulation of Li due to an efficient pump with little leakage would diminish the K concentration of these cells. In this way, Li could prevent the biosynthesis of specific proteins. However, the effect may be reversed by increasing the K supply in the cells. In addition, Li may directly affect the protein biosynthesis apparatus. With sycamore (*Acer pseudoplatanus*) (121), it was concluded that Li did not cause a direct deactivation of the solute uptake machinery during transport. Instead, Li prevented the establishment of the solute uptake mechanism by somehow inhibiting transcription and translation. Thus, the effect of Li was thought to be primarily at the level of protein biosynthesis.

Lithium affects a number of whole plant and cellular processes. However, it has not been possible to determine which effects are primary, and which arise as a consequence of secondary actions due to the complexity of metabolic reactions. Enzymes, necessary components of biochemical reactions, respond to Li levels in the tissue. An examination of Li involvement in enzyme action may account for the variety of changes induced by Li. However, many changes are temporary,

TABLE 3.3.

Observed injured species	Injury symptoms	Approximate time of injury			
		May	Jun	Jul	Aug
apple	tip, then marginal necrosis			X	
Malus sylvestris					
arbor vitae	tip necrosis, then abscission				X
Thuja sp.					
ash; green	marginal chlorosis				X
Fraxinus pennsylvanica			X		
azalea	tip, then marginal chlorosis; then necrosis				
Rhododendron sp.					
bean; bush	marginal chlorosis; then necrosis		X		
Phaseolus vulgaris					
blackberry	tip necrosis and chlorosis	X			
Rubus sp.					
blueberry	tip necrosis		X		
Vaccinium sp.					
broomsedge	tip chlorosis, then necrosis		X		
Andropogon virginicus					
catalpa	marginal chlorosis, then necrosis		X		
Catalpa speciosa					
cattail	tip chlorosis, then necrosis			X	
Typha latifolia					
cedar; eastern red	tip necrosis, then abscission				X
Juniperus virginiana					
cherry; black	marginal necrosis and reddening	X			
Prunus serotina					
corn; field	tip and marginal chlorosis, then necrosis		X		
Zea mays					
corn; sweet	necrosis, tip and marginal chlorosis, then necrosis		X		
Zea mays					
crape myrtle	tip chlorosis, then tip and marginal necrosis		X		
Lagerstroemia indica					
cucumber	marginal chlorosis, then necrosis		X		
Cucurbita sativus					
dogwood	tip necrosis		X		
Cornus sp.					
false solomon's seal	marginal and tip necrosis		X		
Smilacina sp.					
fern	marginal chlorosis, then necrosis			X	
Pteridium sp.					
gladiolus	tip necrosis		X		
Gladiolus sp.					
grape	marginal chlorosis and necrosis			X	
Vitis sp.					
greenbriar	tip and marginal reddening, then necrosis	X			
Smilax sp.					
gum; black	marginal reddening, cupping, then necrosis		X		
Nyssa sylvatica					
gum; sweet	tip chlorosis, then necrosis; some misshapen leaves				X
Liquidambar styraciflua					

TABLE 3.3. (*Continued*)

Observed injured species	Injury symptoms	Approximate time of injury			
		May	Jun	Jul	Aug
hickory; mockernut	marginal chlorosis, then necrosis and	X			
Carya tomentosa	cupping				
hickory; pale	marginal chlorosis, then necrosis			X	
Carya pallida					
hickory; pignut	tip, then marginal chlorosis			X	
Carya glabra					
hickory; shagbark	slight marginal chlorosis			X	
Carya ovata					
holly	tip chlorosis, then tip and marginal				X
Ilex sp.	necrosis				
honeysuckle	chlorosis followed by marginal necrosis			X	
Lonicera sp.					
hydrangea	tip then marginal chlorosis followed		X		
Hydrangea sp.					
iris	tip necrosis		X		
Iris sp.					
johnson grass	tip and marginal necrosis		X		
Sorgum halepense					
juniper	tip necrosis				X
Juniperus sp.					
laurel	tip necrosis		X		
Kalmia latifolia					
liriope	tip necrosis			X	
Liriope muscari					
locust; black	marginal chlorosis; then necrosis and				X
Robinia pseudo-acacia	abscission				
magnolia	tip chlorosis, then tip and marginal			X	
Magnolia grandiflora	necrosis				
maple; red	tip chlorosis, then necrosis	X			
Acer rubra					
maple; silver	tip chlorosis, then necrosis		X		
Acer saccharinum					
maple; sugar	tip chlorosis, then necrosis			X	
Acer saccharum					
morning glory	marginal chlorosis		X		
Ipomoia purpurea					
mulberry	tip and marginal necrosis			X	
Morus rubra					
nandina	tip chlorosis followed by necrosis			X	
Nandina domestica					
oak; blackjack	tip necrosis				X
Quercus marilandica					
oak; chestnut	lobe tip necrosis; cupping	X			
Quercus prinus					
oak; northern red	tip chlorosis followed by necrosis		X		
Quercus rubra					
oak; post	tip necrosis				X
Quercus stellata					

TABLE 3.3. (*Continued*)

Observed injured species	Injury symptoms	Approximate time of injury			
		May	Jun	Jul	Aug
oak; southern red *Quercus saleata*	tip chlorosis, then necrosis	X			
oak; scarlet *Quercus coccinea*	tip necrosis		X		
oak; water *Quercus nigra*	tip necrosis		X		
oak; white *Querbus alba*	lobe tip, then marginal necrosis		X		
oak; willow *Quercus phellos*	tip necrosis		X		
peach *Prunus persica*	tip, then marginal chlorosis and necrosis		X		
pear *Pyrus communis*	marginal chlorosis, then necrosis		X		
peony *Paeonia* sp.	marginal chlorosis, then necrosis	X			
pepper; bell *Capsicum frutesceans*	marginal chlorosis, then necrosis		X		
persimmon *Diospyros virginica*	marginal reddening followed by necrosis			X	
petunia *Petunia* sp.	no observed injury				
pine; loblolly *Pinus taeda*	tip necrosis; leaf abscission		X		
pine; virginia *Pinus virginica*	some juvenile needles; tip necrosis				X
pine; white *Pinus strobus*	tip necrosis		X		
poison ivy *Rhus radicans*	marginal reddening followed by necrosis		X		
pokeberry *Phytolacca* sp.	tip and marginal necrosis		X		
privet *Ligustrum* sp.	marginal chlorosis, then nercrosis			X	
redbud *Cercis canadensis*	tip then marginal chlorosis and necrosis		X		
rhododendron *Rhododendron* sp.	tip chlorosis, then necrosis				X
river cane *Arundinaria tecta*	tip chlorosis, then necrosis		X		
rose *Rosa odorata*	tip and marginal necrosis		X		
rose-of-sharon *Hibicas syriacus*	tip and marginal chlorosis, then necrosis		X		
sassafras *Sassafras albidum*	tip chlorosis then necrosis		X		
sourwood *Oxydendrum arboreum*	tip necrosis; marginal reddening and necrosis			X	

TABLE 3.3. (*Continued*)

Observed injured species	Injury symptoms	Approximate time of injury			
		May	Jun	Jul	Aug
soybean	marginal chlorosis, then necrosis		X		
Glycine max					
spirea	marginal chlorosis, then necrosis; no		X		
Spiraea alba	blooms				
squash	marginal chlorosis, then necrosis	X			
Cucurbita maxima					
strawberry	leaf tooth tip necrosis	X			
Fragaria sp.					
sumac	marginal chlorosis and tip necrosis		X		
Rhus sp.					
sunflower	tip and marginal chlorosis, then necrosis			X	
Helianthus sp.					
sweet pea	tip chlorosis, then necrosis			X	
Lathyrus odoratus					
sycamore	tip necrosis			X	
Platanus occidentalis					
tobacco	marginal and interveinal chlorosis, then		X		
Nicotiana tobacum	marginal necrosis				
tomato	tip chlorosis, then necrosis			X	
Lycopersicon lycopersicum					
tree-of-heaven	marginal chlorosis, then necrosis and		X		
Ailanthus altissima	abscission				
tulip poplar	tip chlorosis, then necrosis			X	
Liriodendron tulipifera					
virginia creeper	marginal and interveinal chlorosis			X	
Parthenocissus sp.					
walnut	marginal chlorosis, then necrosis and leaf		X		
Juglans sp.	abscission				
watermelon	marginal chlorosis, then necrosis		X		
Citrullus vulgaris					
willow; black	marginal chlorosis, then necrosis			X	
Salix nigra					
willow; weeping	marginal chlorosis, then necrosis			X	
Salix babylonica					
wisteria	marginal chlorosis, then necrosis		X		
Wisteria sp.					

These plants from oak-hickory forests, adjacent cultivated fields, road sides, yards, and gardens in the southeast United States were subjected to lithium simultaneously from the air and the soil in which they were growing. The noted responses include the approximate time the first effect was observed, and the symptoms noted. The data represents means of 20 years of observations.

such as those involving turgor. The original premise holds true. Lithium has an effect at the membrane level initially, and subsequently at the metabolic level, either in a primary or secondary way. Evidence presented here would seem to support both possibilities.

Lithium in the Ecosystem

In natural areas adjacent to sources which emit Li into the atmosphere Li deposition may be greater than the rate at which Li is leached from the soil. Hence, soil levels of Li may tend to increase with time. This phenomenon is similar to that described for naval oranges in California (31) where irrigation water containing Li gradually caused soil buildup of Li. These soil levels alone may be high enough to induce injury, if the data in the sensitivity chart (Table 3.1) is accurate, and relevant to the species listed in Table 3.3.

Studies using open-top chambers (122), which can remove most airborne Li with a charcoal filter, indicate that soil alone can be detrimental to vegetation (69). However, this study indicated that air alone (accomplished by placing plants in pots in noncharcoal filtered chambers), can cause typical Li toxicity symptoms. In the natural ecosystem near a lithium-emitting source, vegetation may be affected by Li from the air as well as the soil (69).

Within an ecosystem not all plants are equally responsive to Li. Studies cited in Table 3.3 indicate great variability. McStay's work with eight varieties of beans (84) is a classic example as well. In the natural ecosystem this difference in sensitivity can cause specific changes to occur in the species composition of forests, yards, gardens, and road sides. In oak-hickory forests (*C. tomentosa*) mockernut hickory is lost after several years of such exposure. The same is true for chestnut oak (*Q. prinus*), black cherry (*P. serotina*), laurel (*Kalmia latifolia*), red maple (*Acer rubra*), and southern red oak (*Quercus stellata*). Peonys (*Paeonia* sp.), pears (*Pyrus communis*), roses (*Rosa odorata*), most beans (*Phaseolus* sp.), squash (*Cucurbita maxima*), and spirea (*Spiraea alba*) may not survive in yards and gardens (69). Under the most severe Li impact in an oak-hickory forest the only trees which remain are sweet gum (*L. styraciflua*) and virginia pine (*P. viriginica*). This is not a hypothetical suggestion, but an observed fact (69).

The data presented in Table 3.3 is not generated from controlled experiments. In fact the actual level of Li in the soil and in the air has not been monitored. What is known is that Li is being emitted into the atmosphere at low levels. Lithium levels in the soil, when measured, relate to those listed in Table 3.1 as causing plant injury. When Li is measured in vegetation around local sources the amounts are equivalent to those reported to cause injury. Airborne Li alone has been shown to cause vegetation injury at very low levels where it is emitted into the environment (69). The point stressed in this discussion is that any source that uses Li or Li products needs to consider the potential for vegetation injury. Further, Li insult to vegetation may come from the soil, water, or the air.

References

1. McStay, N.G. (1980). *Effects of lithium on several plant systems*. M.S. Thesis. Department of Botany, North Carolina State University, Raleigh, NC.
2. Weeks, M.E. (1956). Discovery of the elements (6th ed.). *J. Chem. Educ.*, Easton, p. 578.
3. Steinkoenig, L.A. (1915). Lithium in soil. *J. Ind. Eng. Chem.*, 7,425–426.
4. Ivanov, D.N. (1954). The content of rare alkali elements in soils. *Pochvovedenie*, pp. 32–45.
5. Ivanov, D.N. (1956). Occurrence of lithium, rubidium, and cesium in the products of contemporary erosion and in soils. *Kora Vyvetrivaniya*, 2,77–84.
6. Kvanov, D.N., & Muratova, V.S. (1955). The distribution of lithium in saline soils. *Tr. Pochv. Inst. Dokuchaeva Akad. Nauk. SSSR*, 44,294–301.
7. Swain, D.J. (1955). The trace element content of soils. *Tech. Commun. Bur. Soil Sci.*, 48,1–157.
8. Bradford, G.R. (1966). Lithium. In H.D. Chapman (Ed.), *Diagnostic criteria for plants and soil* (p. 793). University of California.
9. Gupta, I.C., Singhla, S.K., & Bharagava, G.P. (1974). Distribution of lithium in some salt affected soil profiles. *J. Indian Soc. Soil Sci.*, 22,88–89.
10. Aldrich, D.G., Buchanan, J.R., & Bradford, G.R. (1955). Effects of soil acidification on vegetation growth and leaf composition of lemon trees in pot culture. *Soil Sci.*, 79,427–439.
11. Bach, R.O., Kamienski, C.W., & Ellestad, R.B. (1967). Lithium and lithium compounds. In R.E. Kirk & D.E. Othmer (Eds.). *Encyclopedia of chemical technology* (2nd ed.), Vol. 12. New York: Interscience Publishers.
12. Bradford, G.R. (1963). Lithium survey of California water resources. *Soil Sci.*, 96,77–81.
13. Gupta, I.C. (1972). Note on lithium in saline ground waters. *Indian J. Agric. Sci.*, 42,650–651.
14. Smith, H.V., Draper, G.E., & Fuller, W.H. (1964). The quality of Arizona irrigation waters. *Ariz. Agric. Exp. Stn. Rep.*, 2234,1–96.
15. Foche, W.O. (1872). Occurrence of lithium in the plant kingdom. *Abh. Naturwiss. Ver. Bremen*, 3,270–275.
16. Focke, W.O. (1878). New observations on lithium in the plant kingdom. *Abh. Naturwiss. Ver. Bremen*, 5,451–452.
17. Robinson, W.O., Steinkoenig, L.A., & Miller, C.F. (1917). The relation of some of the rarer elements in soils and plants. *U.S. Dept. Agr. Bull.*, 600,1–25.
18. Tschermak, E. (1899). The distribution of lithium in plants. *Z. Landwirtsch. Versuchswes. Dtsch. Oesterr.*, 2,560–571.
19. Bertrand, D. (1943). The distribution of lithium in plants. *C.R. Hebd. Seances Acad. Sci.*, 217,707–708.
20. Bertrand, D. (1952). The distribution of lithium in phanerogams. *C.R. Hebd. Seances Acad. Sci.*, 234,2102–2104.
21. Bertrand, D. (1959). Lithium content of seed. *C.R. Hebd. Seances Acad. Sci.*, 249,331–332.
22. Bertrand, D. (1959). New investigations on the distribution of lithium in phanerogams. *C.R. Hebd. Seances Acad. Sci.*, 249,787–788.
23. Bertrand, D. (1959). The influence of altitude on the lithium content of phanerogams plants. *C.R. Hebd. Seances Acad. Sci.*, 249,844–845.

24. Collander, R. (1941). Selective absorption of cations by higher plants. *Plant Physiol.*, *16*,691–720.
25. Yamagata, N., & Takahashi, K. (1951). Absorption of rare alkali metals by plants. *Nippon Kagaku Zasshi.*, *72*,944–947.
26. Ezdakova, L.A. (1964). Lithium in plants. *Bot Zh.* (Leningrad), *49*,1798–1800.
27. Romney, E.M., Wallace, A., Kinnear, J., & Alexander, G.V. (1977). Frequency distribution of lithium in leaves of *Lycium andersonii. Commun. Soil Sci. Plant Anal.*, *8*,799–802.
28. Wallace, A., Romney, E.M., Cha, J.W., & Alexander, G.V. (1974). Sodium relations in desert plants. III. Cation-anion relationships in three species which accumulate high levels of cations in leaves. *Soil Sci.*, *118*,397–400.
29. Wallace, A., Romney, E.M., & Hale, V.Q. (1973). Sodium relations in desert plants. I. Cation contents of some plant species from the Mojave and Great Basin deserts. *Soil Sci.*, *115*,284–287.
30. Cannon, H.L. (1971). The use of plant indicators in ground water surveys, geologic mapping, and mineral prospecting. *Taxon*, *20*,227–256.
31. Aldrich, D.G., Vanselow, A.P., & Bradford, G.R. (1974). Lithium toxicity in citrus. *Soil Sci.*, *71*,291–295.
32. Hilgeman, R.H., Fuller, W.H., True, L.F., Sharples, G.C., & Smith, P.F. (1970). Lithium toxicity in 'Marsh' grapefruit in Arizona. *J. Am. Soc. Hort. Sci.*, *95*,248–251.
33. United States Environ. Prot. Agency, Office of Pesticides and Toxic Substances, TSCA Chemical Assessment Series, Chemical Hazard Information Profiles, August 1976–August 1978 (1980), 1–289.
34. Nobbe, F., Schroeder, J., & Erdmann, R. (1871). On the action of potassium in vegetation. *Landwirtsch. Vers. Stn.*, *13*,321–423.
35. Gaunersdorfer, J. (1887). Plant suppression by specific poisoning with lithium salts. *Landwirtsch. Vers. Stn.*, *34*,171–206.
36. Voelcker, J.A. (1900). The Woburn Pot-Culture Station. A. The Hills' experiments. *J.R. Agric. Soc. Engl.*, *61*,553–591.
37. Voelcker, J.A. (1901). The Woburn Pot-Culture Experiments. I. Pot-culture experiments of 1900. *J.R. Agric. Soc. Engl.*, *62*,317–334.
38. Voelcker, J.A. (1902). The Woburn Experimental Station of the Royal Agricultural Society of England. III. Field experiments, 1901. *J.R. Agric. Soc. Engl.*, *63*,346–361.
39. Voelcker, J.A. (1904). The Woburn Experimental Station of the Royal Agricultural Society of England. II. Pot culture experiments, 1903. *J.R. Agric. Soc. Engl.*, *65*,306–315.
40. Voelcker, J.A. (1910). The Woburn Experimental Station of the Royal Agricultural Society of England. Pot culture experiments of 1909. *J.R. Agric. Soc. Engl.*, *71*,314–325.
41. Voelcker, J.A. (1912). Pot culture experiments, 1910-11-12. I. Hills' experiments. *J.R. Agric. Soc. Engl.*, *73*,314–325.
42. Voelcker, J.A. (1913). The Woburn Experimental Station of the Royal Agricultural Society of England. Pot culture experiments, 1913. I. Hills' experiments. *J.R. Agric. Soc. Engl.*, *74*,411–422.
43. Ravenna, C., & Maugini, A. (1912). The behavior of plants toward lithium salts. *Atti Accad. Naz. Lincei Cl. Sci. Fis. Mat. Nat. Rend.*, *21*,292–298.
44. Ravenna, C., & Zamorani, M. (1909). The behavior of plants toward lithium salts. *Atti Accad. Naz. Lincei Cl. Sci. Fis. Mat. Nat. Rend.*, *18*,626–630.

45. Petri, L. (1910). Observations on the deleterious effects of toxic substances on the olive tree. *Zentralbl. Bacteriol. Parasitenkd. Infectionskr. Hyg. Abt. 2 Naturwiss. Allg. Landwirtsch. Tech. Microbiol.*, *28*,153–159.

46. Hahn, P.D. (1912). Can lithia be a constituent of plant food? *S. Afr. J. Sci.*, *12*,227–229.

47. Brenchley, W.E. (1832). The action on the growth of crops on small percentages of certain metallic compounds when applied with ordinary fertilizers. *J. Agric. Sci.*, *22*,704–735.

48. Eisenmenger, W.S., & Kucinski, K.J. (1940). Minerals in nutrition. II. The absorption by food plants of certain chemical elements important in human physiology and nutrition. *Mass. Agric. Exp. Stn. Res. Bull.*, *374*,12–15.

49. Frerking, H. (1915). The poisonous effect of lithium salts on plants. *Flora* (Jena, 1818–1965), *108*,449–453.

50. Haas, A.R.C. (1929). Mottle-leaf in citrus artificially induced by lithium. *Bot. Gaz.*, *87*,630–641.

51. Epstein, E. (1960). Calcium-lithium competition in absorption by plant roots. *Nature* (London), *185*,705–706.

52. Scharrer, K., & Schropp, W. (1933). Sand and water culture with lithium and rubidium especially regarding their eventual replacement of potassium. *Ernaehr. Pflanze.*, *29*,413–425.

53. Scharrer, K. (1937). The action of ions of the alkali group on the growth of plants, especially the simultaneous influence of potassium and sodium ions. *Forschungsdienst*, *6*,180–187.

54. Kabanov, V.V., & Myasoedov, N.A. (1974). Toxicity of alkaline cations for tomato plants. *Fiziol. Rast.* (Moscow), *21*,391–397.

55. Bingham, F.T., Bradford, G.R., & Page, A.L. (1964). Toxicity of lithium. *Calif. Agric.*, *18*,6–7.

56. Bingham, F.T., Page, A.L., & Bradford, G.R. (1964). Tolerance of plants to lithium. *Soil Sci.*, *98*,4–8.

57. Gupta, I.C. (1974). Lithium tolerance of wheat, barley, rice and gram at germination and seedling stage. *Indian J. Agric. Res.*, *8*,103–107.

58. Rankin, W.H. (1917). The penetration of foreign substances into trees. *Phytopathology*, *7*,5–13.

59. Rumbold, C. (1920). Giving medicine to trees. *Am. For.*, *26*,359–362.

60. Pirschle, K. (1934). Research on the physiological effect of the elements, as shown by growth experiments with *Aspergillus niger* (stimulation and toxicity). *Planta.*, *23*,177–224.

61. Darby, J.F., & Westgatge, P.J. (1958). Lithium as a fungicide on celery. *Proc. Fla. State Hort. Soc.*, *71*,59–62.

62. Kent, N.L. (1941). The influence of lithium salts on certain cultivated plants and their parasitic diseases. *Ann. Appl. Biol.*, *28*,189–209.

63. Vidali, A. (1951). Field experiments with lithium carbonate for control of tobacco mildew. *Not. Mal. Piante*, *16*,35–39.

64. Wortley, W.R.S. (1936). Report of research, 1934-6. The effect of salts of lithium on the resistance of certain plants to disease. *J.R. Agric. Soc. Engl.*, *97*,492–498.

65. Takamatsu, S., Ishizaki, H., & Kunoh, H. (1979). Cytological studies of early stages of powdery mildew in barley and wheat. VI. Antagonistic effects of calcium and lithium on the infection of coleoptiles of barley by *Erysiphe graminis hordei*. *Can. J. Bot.*, *57*,408–412.

66. Wallace, A., Romney, E.M., & Kinnear, J. (1977). Frequency distribution of several trace metals in 72 corn plants grown together in contaminated soil in the greenhouse. *Commu. Soil Sci. Plant Anal., 8*,693-697.

67. Wallace, A., & Romney, E.M. (1977). Synergistic trace metal effects in plants. *Commun. Soil Sci. Plant Anal., 8*,773-780.

68. Wallace, A., Romney, E.M., Cha, J.W., & Chaudry, F.W. (1977). Lithium toxicity in plants. *Commun. Soil Sci. Plant Anal., 8*,773-780.

69. Anderson, C.E. (1989). Unpublished data.

70. Wallace, A. (1979). Excess trace metal effects on calcium distribution in plants. *Commun. Soil Sci. Plant Anal., 10*,473-479.

71. Einor, L.O., & Dzyubak, O.I. (1966). Effect of inorganic salts and organic solvents on the activity of the Hill's reaction with pea chloroplasts. *Ukr. Bot. Zh., 23*,3-10.

72. El-Sheikh, A.M., Ulrich, A., & Boyer, T.C. (1971). Effects of lithium on growth, salt absorption, and chemical composition of sugar beet plants. *Agron. J., 63*,755-758.

73. Sneva, F.A. (1979). Lithium toxicity in seedlings of three cool season grasses. *Plant Soil, 53*,219-224.

74. Rehab, R.I., & Wallace, A. (1978). Excess trace metal effects on cotton. IV. Chromium and lithium in Yolo loam soil. *Commun. Soil Sci. Plant Anal., 9*,645-651.

75. Rehab, F.I., & Wallace, A. (1978). Excess trace metal effects on cotton. III. Chromium and lithium in solution. *Commun. Soil Sci. Plant Anal., 9*,637-644.

76. Edwards, J.K. (1941). Cytological studies of toxicity in meristem cells of roots of *Zea mays*. II. The effects of lithium chloride. *Proc. S.D. Acad. Sci., 21*,65-67.

77. Furuta, T., Martin, W.C., & Perry, F. (1963). Lithium toxicity as a cause of leaf scorch on Easter lily. *Proc. Am. Soc. Hort. Sci., 83*,803-807.

78. Wallihan, E.F., Sharpless, R.G., & Printy, W.L. (1978). Cumulative toxic effects of boron, lithium, and sodium on water used for hydroponic production of tomatoes. *J. Am. Soc. Hort. Sci., 103*,14-16.

79. Nakamura, N. (1904). Can lithium and cesium salts exert any stimulant action on phanerogams? *Bull. Coll. Agric. Tokyo Imp. Univ., 6*,153-157.

80. Hance, F.E. (1933). Chemistry. Hawaiian Sugar Planters' Assoc. Proc. of 53rd Annual Meeting, pp. 46-55.

81. Puccini, G. (1957). Stimulation action of lithium salts on the flower production of the perpetual carnation of the Riveria. *Ann. Sper. Agrar., 11*,41-63.

82. Okhrimenko, M.J., & Kuz'menko, L.M. (1975). The effect of lithium compounds and their importance in plants. In P.A. Vlasyuk (Ed.), *Fertilizers and preparations containing trace elements* (p. 200). Naukova Dumka. Kiev.

83. Vlasyuk, P.A., Okhrimenko, M.F., Sivak, L.A., & Kuz'menko, L.M. (1978). The effect of carboammophoska enriched in lithium on carbohydrate metabolism and productivity of potato. *Agrokhimya., 7*,75-80.

84. McStay, N.G., Rodgers, H.H., & Anderson, C.E. (1980). Effects of lithium on *Phaseolus vulgaris* L. *Sci. of the Total Environ., 16*,185-191.

85. Kent, N.L. (1941). Absorption, translocation, and ultimate fate of lithium in the wheat plant. *New Phytol., 40*,291-298.

86. Birch-Hirschfeld, L. (1920). Investigation of the speed of diffusion of soluble dissolved substances in plants. *Jahrb. Wiss. Bot., 59*,170-262.

87. Hinz, U., & Fischer, H. (1976). Transport of lithium and cesium along the stolons of *Saxifraga sarmentosa* L.Z. *Pflanzenphysiol., 78*,283-292.

88. Jacobson, L., Moore, D.P., & Hannapel, R.J. (1960). Role of calcium in absorption on monovalent cations. *Plant Physiol., 35*,352-357.

89. Laties, G.G. (1959). The development and control of coexisting respiratory systems in slices of chicory root. *Arch. Biochem. Biophys.*, *79*,378–391.
90. Kandeler, R. (1970). The effect of lithium and ADP on the phytochrome regulation of flowering. *Planta.*, *90*,203–207.
91. Englemann, W. (1972). Lithium slows down the *Kalanchoe* clock, *Z. Naturforsch. B: Anorg. Chem. Org. Chem. Biochem. Biophys. Biol.*, *27*,477–478.
92. Englemann, W. (1973). A slowing down of circadian rhythms by lithium ions. *Z. Naturforsch. C: Biochem. Biophys. Biol. Virol.*, *28*,733–736.
93. Desbiez, M.O., & Thellier, M. (1975). Lithium inhibition of the mechanically induced precedence between cotyledonary buds. *Plant Sci. Lett.*, *4*,315–321.
94. Desbiez, M.O., & Thellier, M. (1977). Induced precedence between cotyledonary buds: Ionic or ouabain treatments and memorization effects. In M. Thellier, et al. (Eds.), *Transmembrane Ion Exchange in Plants*, (p. 607). CNRS. Paris.
95. Desbiez, M.O., & Thellier, M. (1978). Ionic control of the occurrence of a biological rhythm for precedence between axillary buds. *Physiol. Veg.*, *16*,785–798.
96. Boyer, N., Chapelle, G., & Gaspar, T. (1979). Lithium inhibition of the thigmomorphogenetic response in *Bryonia dioica*. *Plant Physiol.*, *63*,1215–1216.
97. Louguet, P., & Thellier, M. (1976). The influence of lithium on the degree of opening and speed of opening and closing of stomata in *Pelargonium hortorum*. *C.R. Hebd. Seances Acad. Sci. Ser. D.*, *282*,2171–2174.
98. Carlier, G., & Thellier, M. (1979). Lithium-perturbation of the induction of a methyl-glucose transport during aging of foliar disks of *Pelargonium zonale* (L.) aiton. *Physiol. Veg.*, *17*,13–26.
99. Vlasyuk, P.A., & Okhrimenko, M.F. (1969). Effect of lithium on the photochemical activity of chloroplasts of tomato and pepper. *Dopov. Akad. Nauk. Ukr. RSR. Ser. B: Geol: Geofiz. Khim. Biol.*, *31*,353–356.
100. Vlasyuk, P.A., Okhrimenko, M.F., & Uyazdovskaya, O.S. (1968). The effect of lithium on the photochemical activity of chloroplasts in potato leaves. *Dokl. Vses. Akad. Skh. Nauk. im. V.I. Lenina.*, *11*,5–7.
101. Ezdakova, L.A. (1962). Effect of lithium top-dressing on photosynthesis and respiration in tobacco leaves. *Naukn. Dokl. Vyssh. Shk. Biol. Nauki.*, *2*,137–142.
102. Vlasyuk, P.A., Okhrimenko, M.F., & Kuz'menko, L.M. (1973). Effects of lithium on the content and composition of organic acids in plants of the Solanaceae family. *Fiziol. Biokhim. Kul't. Rast.*, *5*,121–124.
103. Vlasyuk, P.A., Okhrimenko, M.F., & Sivak, L.A. (1976). Effect of lithium on activity of phosphorylase in tomato and potato plants. *Fiziol. Biokhim. Kul't. Rast.*, *8*,493–496.
104. Neskovic, B.A. (1976). New information on the biological effect of lithium. *Period. Biol.*, *78*,148–152.
105. Vlasyuk, P.A., Kuz'menko, L.M., & Okhrimenko, M.F. (1979). The role of lithium in protein-nucleic acid metabolism in plants. *Fiziol. Biokhim. Kul't. Rast.*, *11*,438–447.
106. Hassan, M.N. (1954). The effect of single salt solutions on the histogenesis of radish seedlings. *Alexandria J. Agric. Res.*, *2*,20–27.
107. Powell, J.T., & Richards, E.G. (1972). Specific effects of lithium on stacking equilibria in polynucleotides. *Acta Biochim. Biophys. Acad. Sci. Hung.*, *281*, 145–151.
108. Vlasyuk, P.A., Okhrimenko, M.F., Kuz'menko, L.M., & Sivak, L.A. (1978). Effect of lithium on formation of amino-acyl-tRNA. *Fiziol. Biokhim. Kul't. Rast.*, *10*, 297–301.

109. Vlasyuk, P.A., & Kuz'menko, L.M. (1975). Metabolic activity of potato plant ribosomes in dependence on their supply with lithium. *Fiziol. Biokhim. Kul't. Rast.*, 7,563–568.

110. Vlasyuk, P.A., Kuz'menko, L.M., & Okhrimenko, M.F. (1975). Content and fractional composition of potato protein and nucleic acids under lithium effect. *Dopov. Akad. Nauk. Ukr. RSR. Ser. B: Geol. Geofiz. Khim. Biol.*, pp. 742–748.

111. Vlasyuk, P.A., Okhrimenko, M.F., & Kuz'menko, L.M. (1975). Fractional and amino acidic compositions of proteins and content of free amino acids in potato under the influence of lithium. *Fiziol. Biokhim. Kul't. Rast.*, 7,115–120.

112. Bhattacharyya, B., & Wolff, J. (1976). Stabilization of microtubules by lithium ion. *Biochem. Biophys. Res. Commun.*, 73,383–390.

113. Johnson, F.N. (Ed.). (1975). *Lithium research and therapy* (p. 569). New York: Academic Press.

114. Bello, J., Haas, D., & Bello, H.R. (1966). Interactions of protein-denaturing salts with model amides. *Biochemistry, 5*,2539–2548.

115. Armbruster, A.M., & Pullman, A. (1974). The effect of cation binding on the rotation barrier of the peptide bond. *FEBS Lett., 49*,18–21.

116. Williams, R.J.T. (1973). The chemistry and biochemistry of lithium. In S. Geershon & B. Shopsin (Eds.), *Lithium. Its role in psychiatric research and treatment* (p. 358). New York: Plenum Press.

117. Stracher, A. (1960). Deuterium exchanges or ribonuclease and oxidized ribonuclease in strong salt solutions. *C.R. Trav. Lab. Carlsberg, 30*,468–481.

118. Evans, H.J., & Sorger, G.J. (1966). Role of mineral elements with emphasis on univalent cations. *Annu. Rev. Plant Physiol., 17*,47–76.

119. Kergosien, Y., Thellier, M., & Desbiez, M.O. (1979). Precedence between axillary buds in *Bidens pilosus* L. Modeling at the macroscopic level in terms of catastrophes or at the microscopic level in terms of a cellular "pump and leak". In P. Delattre & M. Thellier (Eds.), *Elaboration and Justification of Models* (p. 343). Paris: Malione.

120. Thellier, M., Desbiez, M.O. (1977). Model of a switching "on" and "off" pump and leak, as a relay and amplification mechanism in the control of morphogenesis. In E. Marre & O. Ciferri (Eds.), *Regulation of cell membrane activities in plants* (p. 332). Amsterdam: Elsevier. North-Holland Biomedical Press.

121. Thellier, M., Thoiron, B., Thoiron, A., Le Guiel, J., & Luttge, U. (1980). Effects of lithium and potassium on recovery of solute uptake capacity of *Acer pseudoplatanus* cells after gas shock. *Physiol. Plant., 49*,93–99.

122. Heagle, A.S., Body, D.E., & Heck, W.W. (1973). An open top field chamber to assess the impact of air pollution. *J. Environ. Qual., 2*,365–368.

4
Transport of the Lithium Ion

Vincent S. Gallicchio

Introduction

Lithium is the lightest metal known, however in nature it exists only as the ion in salts and ores. Lithium has been used medicinally with varying degrees of effectiveness or noneffectiveness in the treatment of a variety of disorders using different chemical forms such as lithium bromide as a hypnotic and sedative agent, lithium chloride as a substitute for table salt, and lithium carbonate in the management of manic depressive disorders (1, 2). The physical and chemical properties of the lithium ion are similar to those of the alkali metal ions, but not in all cases, and in some respects the lithium ion can act similar to the magnesium ion. This is possible because of the location of lithium within the group of alkali metals in the periodic table. The similarity to magnesium includes the high solubility of the halides (except fluoride) in both water and polar organic solvents and the high solubility of the alkyls in hydrocarbons; the low aqueous solubility of the carbonate, phosphate, fluoride, and oxalate; the thermal instability of the carbonate and nitrate; the formation of the carbide and nitride by direct combination; and the reaction with oxygen to form the normal oxide. These properties have been more extensively detailed elsewhere (3, 4). Lithium is tolerated in considerable amounts in biological systems even though lithium is not a normal constituent of biological tissues except in rare minute quantities (5). The presence of lithium in biological tissues therefore may influence or alter the functions of other cations by simply replacing them or by effectively competing for their physiological site of action.

An important aspect of the biology of cations is their distribution amongst biological tissues specifically between the intracellular and extracellular space. This distribution occurs unequally and as a consequence serves as the foundation for the formation of membrane potentials. Therefore the study of cations has focused on the establishment of such membrane potentials based upon the cation concentration gradients in living systems. The involvement of membranes and the transport of cations across membranes serves as the main discussion topic of this chapter; however cations are known to be very effective stimulators of enzyme systems. For example, lithium has been demonstrated to influence the formation and metabolism of biogenic amines which involve enzyme systems that activate receptor association with a specific guanine-nucleotide binding protein (G protein). These G proteins appear to play an important role in postreceptor information transduction and also to influence the action of adenylate cyclase-cyclic AMP systems. The subject of the action of lithium on adenylate cyclase-cyclic AMP systems and activation via phosphoinositol metabolism is covered elsewhere in this text. (See Chapters 7, 8, and 9 in this text). Because lithium has a very narrow therapeutic window its activity as well as its toxicity are a direct function of its absorption by tissues which essentially involves utilized transport mechanisms of this ion into cells. A clearer understanding of these lithium transport mechanisms may provide the answers to understanding not only the clinical utility of lithium in manic depressive episodes, but also the immunopotentiating and antiviral effects of the ion.

Lithium Transport in Red Blood Cells

Numerous investigations on the transport of the lithium ion using human red cells have established that several systems function to mediate lithium transport across the cell membrane. In the presence of ouabain, an effective inhibitor of the Na,K-ATPase, lithium can be transported by a phloretin (3-4-hydroxphenyl-1-[2,4,6,-trihydroxyphenyl]-11-propanone) sensitive sodium-lithium countertransport system (6–8), lithium bicarbonate ion pairs are transported by an anion-exchange system (9) and by the choline exchange system (10). The ouabian (a cardiac glycoside digitalis type vasodepressant) sensitive sodium-potassium pump is also effective in transferring lithium to red cells and at the same time promotes the active efflux of sodium in the absence of any external derived influence of potassium (10). This is accomplished because ouabain is in effect a sodium pump antagonist by its ability to bind to an extracellular site on the cell membrane specific for cardiac glycosides. Binding of the sodium-potassium pump inhibits the transport of sodium out of cells and the transport of potassium into cells. From inside cells lithium can be transported out from the cell and promote potassium influx through the sodium-potassium pump in the absence of internal sodium (12). Further studies have indicated that lithium can replace sodium in the ouabain insensitive sodium-potassium exchange system (13, 14). This transport system appears to have a higher affinity for lithium than for sodium and also indicates this system may differ within every individual with regard to the maximum rate of transport. For example, it has been demonstrated that in individuals with hypertension and their families this transport mechanism may be genetically determined. Also there exists an ouabain insensitive transport mechanism for lithium ions where lithium ions can effectively replace sodium but not potassium on the outward sodium-potassium cotransport system in human red cells (15). There exist two ouabain insensitive systems where the lithium-potassium cotransport system has an apparent affinity for lithium that is one-half that for sodium and 30 times lower than the lithium-sodium countertransport system.

Many investigations have focused on the human erythrocyte as a model cell system for studying membrane transport processes from patients with various

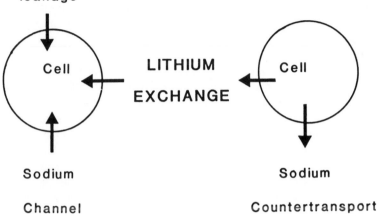

FIGURE 4.1. Paths by which lithium crosses cell membranes.

psychiatric disorders. This system has been advocated because it had been postulated that the transport of lithium across red cell membranes from these individuals has been demonstrated to be altered when compared to red cells of normal controls. In manic-depressive patients treated with lithium, two slowly developing changes have been reported in cation distributions across the red cell membrane: these have been an inhibition of the lithium/sodium countertransport by about 50% and an increase in red cell intracellular choline by a factor of ten or greater (16–19). These two transport mechanisms are of interest because of their possible relevance to the mechanism of lithium therapy. This relationship exists because of the observation that the appearance of these transport effects coincide with the rise of plasma lithium. In addition, the inhibition of the countertransport system develops with a half-time of several days and reverses within two weeks following the cessation of lithium therapy. Furthermore, the lithium effect on inhibiting choline transport is present long after the lithium levels in both red cells and plasma have returned to their normal baseline values indicating there also may be an age-related association with lithium transport in young red cells compared to aged red cells (20).

The importance of lithium transport into cells is that a condition must exist before lithium may induce its therapeutic activity. The steady-state distribution of lithium between red cells and the blood plasma is commonly referred to as the lithium ratio and has been related to the clinical response of lithium therapy, clinical diagnosis, and side-effects (see Figure 4.1). This subject is more extensively detailed in reference 5. In general, manic depressed patients express higher lithium ratios (cell concentration versus blood plasma) than patients who do not respond to lithium therapy. All investigations have generally indicated that there is a large interindividual variation in the lithium ratio amongst patients and normal controls. These steady-state levels of lithium in red cells are usually considerably below the levels measured in plasma thus suggesting that the intracellular lithium level would be a more significant value than the plasma level in determining lithium concentrations in the brain of treated subjects. The majority of studies have indicated the lithium distribution ratio amongst treated patients is less than 1.0 indicating a lithium transport mechanism must act across the red cell membrane to keep this ratio below 1.0.

The time course and age relationships to lithium transport, especially via lithium-sodium countertransport inhibition, are significant clinically because the action of lithium may be intracellular and countertransport is the process maintaining therapeutic lithium levels within red cells, nerve cells and muscle cells. This is important because nerve cells are the site of lithium's therapeutic action and muscle cells are the site of the toxic effects of lithium. Since the therapeutic effects of lithium usually take a week to develop to control manic depression and the toxic effects of lithium also are slow to develop, this indicates the inhibition of the countertransport mechanism may be relevant to the slow therapeutic action of lithium.

Lithium Transport in Fibroblasts

Lithium has been proven to be an effective treatment for manic depressive illness and since the pathogenesis and/or the treatment of this disorder may be related to a central nervous system membrane difficulty of lithium transport, alternative cell systems have been utilized to investigate the mechanism of lithium action. These alternative cell systems have used fibroblasts as the cell studied. This system has been chosen by investigators because it provides a cell system that is both readily accessible in human subjects and reflects the transport of lithium by central nervous cells (21). Studies have determined the lithium distribution ratio in cultured skin fibroblasts differs significantly from the ratio in erythrocytes (22). More recent studies have compared cultured fibroblasts prepared from normal controls, and patients afflicted with either schizophrenia or manic depression for their 24-hour lithium ratios and steady-state membrane potential (23). No significant difference was observed in lithium uptake from any group, nor was there any difference between fibroblast uptake versus uptake when erythrocytes from the same groups were analyzed. These authors concluded that there was no correlation between the lithium ratio and membrane potential.

Previous studies using red blood cells have shown that erythrocytes would serve as a useful model system to study central nervous system membrane function in lithium transport. These studies found a high correlation between the lithium ratio and such measures of lithium transport (17, 22). Bipolar patients in general have a high lithium ratio; however other investigations have been unable to replicate these findings in determining the lithium ratio (24, 25). Therefore the use of fibroblasts appears to eliminate many of the sources of variability to similar measurements made using erythrocytes. These studies have concluded that cells from manic-depressive patients do not differ in the lithium ratio when compared to medication-free controls. Additional studies incorporating fibroblasts have demonstrated lithium may effectively use a transport mechanism that is sensitive to the transport of sodium because lithium transport is completely blocked by the sodium transport inhibitor amiloride (3,5-diamizo-N-[aminoimino-methyl]-6-chloropyrazinecarboxamide) (26). This observation will be discussed in a later section referring to further blood cell effects of lithium.

Lithium Transport in Transformed Neuroblastoma Cells

Because lithium is presumed to exert its psychotherapeutic action within or associated with brain tissue, and because the direct study of human brain tissue for examination obviously is difficult, the use of erythrocytes as an alternative cell system for study was adapted. However, as was indicated previously, the red cell is not capable of exerting an action potential and therefore may not serve as an ideal model cell especially since the activation of the action potential involves the presence and unique properties of the sodium channel.

Neuroblastoma cells have been studied to provide a cell type that expresses many morphological biochemical and physiological properties of neurons (27). The results of studies on the transport of lithium using cultured neuroblastoma cells suggest that in electrically excitable tissue the distribution of lithium depends upon the unimpaired condition of the cell membrane (28). Results of these studies imply that the red cell, despite having the advantages of accessibility and sampling convenience, may not serve as the best model to understand the transport, regulation, and distribution of lithium in the brain.

Lithium-Induced Leukocytosis and Transport Mechanisms

A well-described side-effect in manic depressive patients receiving lithium as therapy is a reversible leukocytosis that is demonstrated by neutrophilia and lymphocytopenia (29). It has been a well-described observation from studies from other cell types that ionic channels can also play an important role in the activation and function of leukocytes (30). The opening of a single or several channels on the surface of leukocytes can, by altering the membrane potential, significantly influence the activity of the cell. Although the hematopoietic effects of lithium have been well described and are treated separately in this text (see Chapter 6) studies have been conducted that indicate special transport mechanisms are involved and therefore they are discussed in this chapter.

Transport of sodium ions in exchange for potassium ions is achieved at the cell membrane by the activation of the NaK-ATPase and this mechanism, often referred to as the sodium pump, at least in part has been implicated as one of the transport mechanisms whereby lithium enters cells. This enzyme is activated by sodium and potassium and is a magnesium-dependent ATPase. This NaK-ATPase, when activated by sodium or potassium ions, stimulates a pumping mechanism that expresses extracellular binding sites that are specific for cardiac glycosides such as ouabain. Ouabain binding effectively blocks the action of the NaK-ATPase and therefore limits the activity of this sodium pump, thereby decreasing intracellular sodium ion concentrations. The binding of ouabain at this external site on the cell surface appears to effectively compete with potassium binding which then activates the pump mechanism (31).

The use of cardiac glycosides and their subsequent effects on cellular proliferation have suggested a role for monovalent cations like lithium in augmenting cellular processes. Specifically, the agent ouabain has been shown to be a potent inhibitor of lymphocyte proliferation, the mechanism of which is attributed to blockage of the Na/K pump (31, 32). This blockage has therefore implicated the transport mechanism of sodium and potassium ions, and thus can influence various physiological functions. The observation that lithium can influence many lymphocyte reactions may in part be attributable to lithium transport processes present in lymphocytes (See Chapter 5).

Data from experiments performed in our laboratory have demonstrated that lithium stimulates the production of granulocytes by mechanisms that involve the utilization of sodium transport pathways. When bone marrow cells were exposed to ouabain either before or following lithium, ouabain produced an irreversible reduction in the ability of lithium to increase the number of granulocyte precursors (34). These studies have concluded that this inhibition of granulocyte precursors in the presence of ouabain may be the result of a direct action on the precursor stem cell responsible for the production of differentiated progeny and not due to an influence on the production of its necessary growth factor, because growth factor production is not affected by ouabain exposure. Therefore these observations suggest that by blocking the NaK-ATPase, ouabain can significantly reduce the ability of lithium to increase granulocyte precursors. It is unlikely that the inhibition of granulocyte precursors is the result of ouabain exposure leading to cell death and therefore just a manifestation of cytotoxicity since at similar concentrations, ouabain effectively stimulates and increases the number of erythroid progenitors leading to increased red cell production (35, 36).

To further investigate the importance of sodium transport pathways in the mechanism whereby lithium increases granulocyte precursors, we conducted further studies incorporating agents known to alter sodium ion permeability. At a noncytotoxic concentration and after various timed exposures, gramicidin and valinomycin, both cyclic polypeptide antibiotics and ionophores, were added to bone marrow cell cultures before the addition of lithium. In the presence of either agent, the ability of lithium to increase granulocyte precursors was not as significantly reduced when compared to lithium control cultures. However, in the presence of gramicidin the number of granulocyte precursors was not as significantly reduced when compared to those cultures plated in the presence of valinomycin, a potassium-specific ionophore.

Since these preliminary studies indicated that sodium permeability may be an important mechanism whereby lithium stimulates granulocyte production, additional studies were conducted using agents that more precisely inhibit sodium transport, namely amiloride and phloretin. Amiloride has been shown to be an effective inhibitor of passive sodium transport in a number of tissues and is used clinically as a potassium-sparing diuretic. Phloretin [2',4',6'-trihydroxy-3-(p-hydroxyphenyl)-propiophenone] is an effective agent for inhibiting sodium ion permeability. In the studies performed in our laboratory amiloride was effective in reducing the increase in granulocyte precursors usually observed in control lithium cultures. The degree of amiloride-induced granulocyte precursor reduction was 15% to 64%. In addition, cultures with phloretin and lithium reduced granulocyte precursors when compared to the number of precursors obtained in lithium control cultures. These results demonstrate a role for sodium transport pathways as a possible mechanism for lithium augmentation of leukocyte stimulation since in the presence of these sodium ion transport inhibitors, a significant reduction in the ability of lithium to increase granulocyte precursors was observed. This reduction, however, was significantly higher than the precursor values obtained in cultures containing only amiloride or phloretin. Lithium was partially effective in reversing this inhibition.

Elevated sodium ion concentrations have also been reported to increase intracellular calcium concentrations by releasing calcium from intracellular storage pools or by reducing calcium efflux by sodium/calcium countertransport. This link between altered sodium-calcium transport has been suggested to be a possible mechanism for the ability of certain cations to modulate blood cell production. Ouabain elevates intracellular sodium ion concentrations which can activate the sodium/calcium countertransport system. This results in the elevation of intracellular calcium ion concentrations. This is a possible mechanism whereby ouabain increases red cell production.

In order to explore this association between sodium transport flux, calcium permeability, and lithium, our laboratory investigated the effect of the antibiotic calcium ionophore, A23187 on the augmentation of granulocyte precursors. These studies demonstrated that in the presence of A23187 and lithium, granulocyte precursors were significantly reduced when compared to lithium controls. This observation indicates calcium may restrict the capability of lithium to increase granulocyte precursors. It is possible that A23187 is acting as a strong lithium ionophore and that by doing so, binds lithium so effectively that lithium's ability to influence granulocyte precursors is significantly reduced. Two molecules of A23187 effectively bind one divalent cation; however in the case of lithium one A23187 molecule binds one lithium monovalent cation. It has been reported that lithium can compete with certain divalent cations such as calcium and magnesium (37). Because calcium also exists in intracellular storage pools, the amount of lithium used in these studies may have been insufficient to reverse the reduction of granulocyte precursors observed in the presence of A23187. This substitution of lithium for certain cations may explain the reported effects of calcium and lithium on in vitro red cell production where calcium has been shown to be an effective stimulator of red cell production (38, 39), while our laboratory has demonstrated lithium reduces red cell production (40, 41). Further studies are required to elucidate the relationships between lithium cation flux and the control of blood cell production.

These studies have implied that monovalent cations and their transport processes may directly influence the direction of a number of enzymatic, cellular, and therefore physiological processes. These effects that may in part be regulated by ions such as lithium may explain its therapeutic action as well as the basic underlying defect leading to an altered physiological state. The regulatory influences of cation transport on cellular proliferation are a current topic of investigation. Although a relationship between membranous cation flux and cellular activation preceeding cell division is generally accepted, the actual mechanism remains unclear and awaits further experimentation.

Analytical Measurements of Lithium

Several analytical methods have been used to measure lithium levels in biological fluids. This methodology has included atomic absorption, flame emission, spectrophotometry, flame emission, ion-selective electrodes, neutron activation analysis, and nuclear magnetic resonance (NMR). The vast majority of these technologies have limitations in their measurements of lithium analysis by assessing only total intracellular or extracellular lithium concentrations within cells. With the advent of NMR techniques, it is now possible to measure and differentiate intracellular pools of lithium contained within cells, i.e., free intracellular lithium from membrane-bound lithium. This capability of measuring different lithium pools will provide for a more complete understanding of both the binding as well as the transport properties of lithium within cells. With the advent of ^7Li NMR methods, this technique will allow for this assessment of the various lithium pools because magnetically or chemically nonequivalent sites of lithium have different chemical and relaxation properties (43). Recent reviews have discussed and compared both the advantages and disadvantages of these conventional methodologies (44, 45).

A basic underlying problem common to conventional lithium measurements is that using blood samples requires centrifugation followed by cell lysis before measuring lithium from either red cells or plasma. Lithium concentrations, present either intracellularly or extracellularly, are then analyzed and compared to standard lithium solutions that are similar in composition to the blood samples under analysis. The methods employed in these conventional analytical techniques produce nonspecific lithium binding to both membranes and intracellular components. Thus an ideal measurement of lithium transport should involve an intact measurement system. The development of a radioactive lithium isotope ^7Li has allowed the technique of lithium analysis by NMR. The ^7Li isotope contains a nucleus with increased sensitivity, thus allowing for its routine detection by NMR spectroscopy. With this technology the only specific instrumentation required is a multinuclear probe that is capable of covering the resonance frequency of the ^7Li nucleus. Using a different frequency one can also detect lithium using another lithium isotope ^6Li; however, the detection system is much more time consuming because of different properties expressed by the ^6Li isotope.

Use of this lithium NMR technology has produced studies in the analysis of lithium transport, both passive and ionophore induced, that involved both red cells and liposome measurements (46–51). One study, utilizing NMR analysis, has compared ion transport from the red cells taken from six bipolar patients prior to lithium treatment and following one week of lithium therapy. Differences observed in these studies have indicated that patients with bipolar mental disorders may have changes in their tissue solvent structure which can be influenced following lithium treatment (52). NMR technology has also provided additional studies designed to identify the various organ/tissue distribution of lithium following both single and multiple injections of lithium carbonate (53); therefore the use of lithium tissue/organ distribution analysis by NMR techniques may continue to improve our understanding of the basic pharmacology of the lithium ion.

References

1. Lithium: Inorganic Pharmacology and Psychiatric Use, Proceedings of the 2nd British Lithium Congress, Wolverhampton, U.K. (1988). N.J. Birch (Ed.). Oxford: IRL Press, pp. 1–340.
2. Schou, M. (1985). Lithium side effects of drugs annual 9 (pp. 27–32). M.N. Duke (Ed.). Amsterdam: Elsevier Science Publisher.
3. Bach, R.O., Ellestad, R.B., Kamienski, C.W., & Wasson, J.R. (1981). Lithium and lithium compounds. *Encyclopedia of chemical technology*, Vol. 14 (pp. 448–476). New York: John Wiley & Sons, Inc.
4. Bach, R.O. (1987). Lithium and viruses. *Medical Hypothesis, 23,*157.
5. Pandey, G.N., & Davis, J.M. (1980). Biology of the lithium ion. In A.H. Rossof & W.A. Robinson (Eds.). *Lithium effects on granulopoiesis and immune function. Advances in Experimental Medicine and Biology.* Vol. 127 (pp. 15–59). New York: Plenum Press.
6. Haas, M., Schooler, J., & Tosteson, D.C. (1972). Coupling of lithium to sodium transport in human red cells. *Nature* (London), *258,*425.
7. Duhm, J., Eisenreid, F., Becker, F., & Greil, W. (1976). Studies on the lithium transport across the red cell membrane. *Pflugers Arch. Eur. J. Physiol., 364,*147.
8. Pandey, G.N., Sarkadi, B., Haas, R., Gunn, R.B., Davis, J.M., & Tosteson, D.C. (1978). Lithium transport pathways in human red blood cells. *J. Gen. Physiol., 72,*233.
9. Funder, J., Tosteson, D.C., & Weith, J.O. (1978). Effects of biocarbonate on lithium transport in human red cells. *J. Gen. Physiol., 71,*721.
10. Martin, K. (1977). Choline Transport in Red Cells. In J.C. Ellory & V.L. Lew (Eds.). *Membrane transport in red cells* (pp. 101–103). London: Academic Press.
11. Beauge, L.A., & Del Campillo, A. (1976). The ATP dependence of a ouabain-sensitive sodium efflux activated by external sodium, potassium and lithium in human red cells. *Biochem. Biophys. Acta, 433,*547.
12. Duhm, J., & Becker, B.F. (1979). Studies on lithium transport across the red cell membrane. *J. Membr. Biol., 51,*263.
13. Pandey, G.N., Baker, J., Chang, S., & Davis, J.M. (1978). Prediction of in vivo red cell/plasma Li+ ratios by in vitro methods. *Clin. Pharmacol. Ther., 24,*343.

14. Sarkadi, B., Alifimoff, J.K., Gunn, R.B., & Tosteson, D.C. (1978). Kinetics and stoichiometry of Na-dependent Li-transport in human red blood cells. *J. Gen. Physiol., 72*,249.

15. Canessa, M., Bize, F.M., Adragna, N., & Tosteson, D.J. (1982). Co transport of lithium and potassium in human red cells. *J. Gen. Physiol., 80*,149.

16. Meltzer, H.L., Kassir, S., Dunner, D.L., & Fieve, R.R. (1982). Repression of a lithium pump as a consequence of lithium ingestion by manic-depressive subjects. *Psychopharm., 54*,113.

17. Rybakowski, J., Frazer, A., & Mendels, L. (1978). Lithium efflux from erythrocytes incubated in vitro during lithium carbonate administration. *Commun. Psychopharmacol., 2*,105.

18. Lee, G., Lingsch, C., Lyle, P.T., & Martin, K. (1974). Lithium treatment strongly inhibits choline transport in human erythrocytes. *Brit. J. Clin. Pharmacol., 1*,365.

19. Lingsch, C., & Martin, K. (1976). An irreversible effect of lithium administration to patients. *Brit. J. Pharmacol., 57*,323.

20. Diamond, J.M., Meier, K., Gosenfield, L.F., Jope, R.S., Jenden, D.J., & Wright, S.M. (1983). Recovery of erythrocytes $Li+/Na+$ counter transport and choline transport from lithium therapy. *J. Psychiat. Res., 17*,385.

21. Mendels, J., & Frazer, A.J. (1973). Intracellular lithium concentration and clinical response towards a membrane theory of depression. *Psychiat. Res., 10*,9.

22. Breslow, R.E., Demuth, G.W., & Weiss, C. (1985). Lithium incorporation in the fibroblasts of manic depressives. *Biol. Psych., 20*,58.

23. Hitzemann, R., Kao, L., Hirschowitz, J., Garver, D., & Gruenstein, E. (1988). Lithium transport in human fibroblasts: Relationship to RBC lithium transport and psychiatric diagnosis. *Psychiat. Res., 24*,337.

24. Szentistvanyi, I., & Janka, Z. (1979). Correlation between the lithium ratio and Na-dependent Li transport of red blood cells during lithium prophylaxis. *Biol. Psychiat., 14*,973.

25. Frazer, A., Mendels, J., & Brunswick, D. (1978). Erythrocyte concentrations of the lithium ion: Clinical correlates and mechanisms of action. *Amer. J. Psychiat., 135*,1065.

26. Mendelewicz, J., & Verbanek, P. (1977). Lithium ratio and clinical response in manic-depressive illness. *Lancet, 1*,41.

27. Smith, J.B., & Rozengurt, E. (1978). Serum stimulates the $Na+$, $K+$ pump in quiescent fibroblasts by increasing Na entry. *PNAS, USA, 75*,5560.

28. Richelson, E. (1977). Lithium ion entry through the sodium channel of cultured mouse neuroblastoma cells: A biochemical study. *Science, 196*,1061.

29. Gorkin, R.A., & Richelson, E. (1981). Lithium transport by mouse neuroblastoma cells. *Neuropharmacol., 20*,791.

30. Gallin, E.K. (1986). Ionic channels in leukocytes. *J. Leuko. Biol., 39*,241.

31. Shopsin, B., Friedmann, R., & Gershon, S. (1971). Lithium and leukocytosis. *Clin. Pharmacol. Ther., 12*,923.

32. Quastel, M.R., & Kaplan, J.G. (1970). Lymphocyte stimulation: The effect of ouabain on nucleic acid and protein synthesis. *Exp. Cell Res., 62*,407.

33. Robinson, J.D., & Flaschner, M.S. (1979). The $(Na+ + K+)$ activated ATPase enzymatic and transport properties. *Biochem. et Biophysica Acta, 549*,145.

34. Spivak, J.L., Misiti, J., Stuart, R., Sharkis, S.J., & Sensenbrenner, L.J. (1980). Suppression and potentiation of mouse hematopoietic progenitor cell proliferation by ouabain. *Blood, 56*,315.

35. Gallicchio, V.S., & Murphy, M.J., Jr. (1982). Erythropoiesis in vitro IV. Ouabain effects on erythroid stem cells. *Stem Cells, 1*,30.
36. Gallicchio, V.S., DellaPuca, R., & Murphy, M.J., Jr. (1982). Effects of ouabain and valinomycin on in vitro colony formation (CFU-E and BFU-E). *Exper. Cell Biol., 50*,295.
37. Birch, N.J. (1974). Lithium and magnesium dependent enzymes. *Lancet, ii*,965.
38. Misiti, J., & Spivak, J.L. (1979). Erythropoiesis in vitro. Role of calcium. *J. Clin. Invest., 64*,1573.
39. Gallicchio, V.S., Chen, M.G., & Murphy, M.J., Jr. (1982). Modulation of murine in vitro erythroid and granulopoietic colony formation by ouabain, digoxin and theophylline. *Exper. Hematol., 10*,682.
40. Gallicchio, V.S., & Chen, M.G. (1980). Modulation of murine pluripotential stem cell proliferation in vitro by lithium carbonate. *Blood, 56*,1150.
41. Gallicchio, V.S., & Chen, M.G. (1981). Lithium stimulates the proliferation of hematopoietic stem cells. *Exper. Hematol., 9*,804.
42. Gallicchio V.S., Murphy, M.G., Jr. (1983). Cation influences on in vitro growth of erythroid stem cells (CFU-E and BFU-E). *Cell Tiss. Res., 733*:175–181.
43. Detellier, C. (1983). Alkai Metals. In P. Laszlo (Ed.). NMR of newly accessible nuclei, Vol. 2 (pp. 105–151). New York: Academic Press.
44. Xie, Y.X., & Christian, G.D. (1987). Measurement of serum lithium levels. In F.N. Johnson (Ed.). *Depression and mania* (pp. 78–88). Oxford: IRL Press.
45. Vartsky, D., LoMonte, A., Ellis, K.J., & Yasumura, S. (1988). A method for *in vivo* measurement of lithium in the body. In N. Birch (Ed.). *Lithium: Inorganic pharmacology and psychiatric use* (pp. 297–298). Oxford: IRL Press.
46. Espanol, M.T., & de Freitas, Mota. (1987). ^{7}Li NMR studies of lithium transport in human erythrocytes. *D. Inorg. Chem, 26*,4326.
47. Pettegrew, J.W., Post, J.F.M., Panchalingham, K., Withers, G., & Woessner, D.E. (1987). ^{7}Li NMR study of normal human erythrocytes. *Magn. Reson., 71*,504.
48. Hughes, M.S., Flavell, K.J., & Birch, N.J. (1988). Transport of lithium into human erythrocytes as studied by ^{7}Li NMR and atomic absorption spectroscopy. *Biochem. Soc. Trans., 16*,827.
49. Ridell, F.G., & Arumugam, S. (1988). Surface charge effects upon membrane transport processes: The effects of surface charge on monensin-mediated transport of lithium ions through phospholipid bilayers studies by ^{7}Li NMR spectroscopy. *Biochim. Biophys. Acta, 945*,65.
50. Shinar, H., & Navon, G. (1986). Novel organometallic ionophore with specificity toward Li+. *J. Amer. Chem. Soc., 108*,5005.
51. Rosenthal, J., Strauss, A., Minkoff, L., & Todd, L.E. (1986). Identifying lithium-response bipolar depressed patients using nuclear magnetic resonance. *Amer. J. Psych., 143*,779.
52. Duhm, J. (1982). Lithium transport pathways in red blood cells. *Excerpta Med.,* 1–20.
53. Renshaw, P.F., & Wicklund, S. (1986). Relaxation and imaging of lithium in vivo. *Biol. Psych. Magn. Reson. Imag., 4*,193.

5
Modulation of Immune System Elements by Lithium

DAVID A. HART

Introduction

Lithium is one of the lighter elements of the alkali metal series. The element was present at the formation of our galaxy (1) and is widely distributed in the earth's crust (0.005% by weight). It is present in a number of minerals and occurs in certain brines mostly as the chloride salts (1a). The element occurs as two natural isotopes, ^6Li (7.6%) and ^7Li (92.4%). The two isotopes can be separated and preparations enriched for either ^6Li (95%) or ^7Li (> 99.9%) can be obtained. No naturally occurring radioactive isotopes of Li exist and those artificially generated are very unstable (T1/2 < 1 second).

While widely distributed in the earth's crust, present in water-soluble form in brines, and obviously present since the formation of the earth, lithium is somewhat of a "biologically neglected" element. In contrast to hydrogen, sodium, potassium, magnesium, calcium, and other trace elements, lithium does not appear to play a normal cofactor function in any known enzyme or transport system. However, Li is well tolerated by most organisms exposed to reasonable concentrations. The element is found in low to modest levels in many water supplies and individuals or animals exposed to the element either naturally or overtly do not appear to suffer from any obvious alterations in the integrity or functioning of biological regulation. However, it is obvious from other chapters of this book as well as earlier publications (2–6), that Li is effective in influencing a number of biological systems both in vitro and in vivo. Therefore, Li may actually have been included in biological regulation as a subtle "regulator" of homeostatic mechanisms and may depend on its previously discussed exclusion from intimate involvement in critical systems in order to perform a role as a general regulator of homeostasis. While this theme is admittedly speculative, results to be discussed in this chapter, as well as possibly others, will, I believe, lend credence to such a possibility.

Historical Perspectives

The early history of the biological and biomedical effects of lithium have been reviewed by Johnson (2, 3) and will not be discussed here. The modern history of the biomedical application of lithium is generally associated with the observations of Cade (4, and reviewed in 2, 3) who reported on the therapeutic efficacy of Li in the treatment of affective disorders. Since this post-World War II "rebirth" of interest in lithium, lithium salts have been used and abused in a number of biomedical circumstances. Over the past forty years it has been successfully used for the treatment of a subset of affective disorders and its efficacy is generally accepted. Efficacy is usually obtained with concentrations yielding plasma values of approximately 1 meq. Higher levels very often lead to toxicity and there seems to be a tenuous balance between therapeutic efficacy and unwanted side effects. A number of organ systems have been demonstrated to be involved in such side effects and these include the kidneys, muscles, endocrine systems, cardiovascular system, and elements of the immune system. Thus, the Li salts are effective in the treatment of affective disorders but can impact on a number of biological systems.

Biochemical Basis for the Therapeutic Efficacy of Lithium Salts

The side effects of lithium described earlier have provided the impetus for many studies directed towards a better understanding of the biochemical basis of Li action. After a plethora of studies, it is probably fair to say that we know much about the biochemical impact of Li on cells and tissues, but still do not have a clear idea of how Li influences affective disorders (5). Thus, in both in vitro and in vivo investigations, Li salts have been shown to influence a number of systems including ion transport systems, cyclic AMP-dependent systems, adenylcyclase, phosphoinositide metabolism (via inhibition of enzymes) as well as other systems (reviewed recently in Birch, 6). Much of the effort has focused on intracellular second messenger systems that can be regulated by intercellular stimuli. While considerable information has been gained regarding the potential for Li to influence specific biochemical steps in biological regulation, it should be pointed out that much of it was obtained with concentrations of Li exceeding therapeutic limits. Thus, we may have still not identified the appropriate "biochemical derangement" relating to Li efficacy in affective disorders, or the degree of derangement is such that concentrations less than those leading to maximal in vitro changes are necessary to correct the alteration in patients. Alternatively, the pleiotrophic effect of Li on second messenger and ion transport systems may be the key to a better understanding of the biochemical basis of Li action. That is, the key to lithium's action may be due to its ability to subtly influence a number of systems rather than just a single target system. Since many second messenger systems are interactive, such a line of thinking may be more appropriate. Such uncertainty may very well be resolved when the genes responsible for such affective illnesses are characterized (7, 8).

Lithium Effects on Animal Behavior

Lithium salts have been investigated for their ability to alter animal behavior as well as human affective behavior (9). Interestingly, some of Cade's early experiments described the effect of Li salts on the behavior of guinea pigs (3). Other investigators have used LiCl in establishing conditioning responses of a number of animal species ranging from fish and snakes to birds and mammals (9). A number of questions remain unanswered with regard to Li effects on animal behavior and several of these are related to test analysis and toxic effects of Li rather than therapeutic effects (9). Therefore, while such studies have not generated conclusive evidence for the ability of Li to influence animal models of behavior disorders, these investigations have prompted further animal studies of lithium's effect on other biological systems.

Impact of Lithium on the Immune System

One of the other biological systems that has been investigated is the immune system. This area of investigation has been prompted by reports describing effects of Li treatment on elements of the immune system in patients with affective disorders (10–12). Many patients experience a leukocytosis while on Li therapy. However, lithium has also been reported to influence the disease activity of a number of autoimmune states such as arthritis, psoriasis, and myasthesia gravis, as well as others (10). In addition, some patients on Li therapy develop autoantibodies in their plasma (12, 13). Such serendipity has led to analysis of Li as a therapeutic modality in immune dysfunction as well as the use of Li as a tool to further understand the immune system. While the latter application is, at first glance, distinct from the use of Li in treating affective disorders, some interesting insights have emerged.

Lithium has been reported to influence a number of immunologically relevant cells and systems both in vivo and in vitro. The cells reported to be susceptible to Li influence include polymorphonuclear leukocytes (PMN), monocyte/macrophages, lymphocytes, suppressor T-cells, NK cells, bone marrow cells, and thymocytes (10–12). Because of the diversity of cells influenced, the following discussion will focus on groups of cells, based on function.

Influence of Lithium on PMN, NK, and Mast Cells

PMN

PMN and PMN-precursors have been found to be influenced by Li both in vivo and in vitro (10–12). Patients receiving Li for affective disorders often experience a leukocytosis. This is apparently an effect on the bone-marrow environment mediated by enhanced release of growth factors which influence PMN differentiation from multipotential stem cells. This effect of Li can be observed in vitro with bone marrow cells (11, 12; Chapter 9, this volume). The effect of Li can be abolished by removal of a phagocytic bone marrow cell, which is presumably the source of the growth factors (16), or by supplementing the cultures with maximal quantities of exogenous growth factors (16). This aspect of lithium's impact on the bone marrow will be discussed in detail by Dr. V. Gallicchio [this volume].

Lithium can also impact on mature PMN function in vivo. Perez et al. (17) have reported that Li treatment of a patient with chemotaxis-defective PMN experienced a complete recovery of PMN function. The PMN from this patient contained elevated levels of cyclic AMP (cAMP) and Li treatment depressed the cAMP levels to the normal range. Interestingly, Li did not influence the chemotaxis of PMN from control individuals. Therefore, Li could "correct" the aberrant PMN function but had minimal effect on normal PMN chemotaxis.

In vitro studies have revealed that Li can enhance degranulation of PMN (18–22). Bloomfield and Young (14, 18) reported that incubation of human PMN with Li salts led to enhanced release of enzymes from both specific and azurophilic PMN granules. Reports from this laboratory indicated that Li could stimulate enzyme release from a subset of specific granules (lysozyme) but did not stimulate elastase release from azurophilic granules (20–22). Interestingly, ^{6}Li was more effective than ^{7}Li in stimulating lysozyme release, particularly when lower concentrations of LiCl were tested.

Incubation of elicited rat PMN with Li led to levels of lysozyme release comparable to those released from human PMN (20). However, it has not been determined if Li was acting on the same subset of granules in PMN from both species.

The finding that Li stimulates exocytosis of a specific subset of granules in human PMN (22) raises the possibility that the intracellular signals regulating this subpopulation are uniquely sensitive to Li influence while those regulating other subpopulations of granules use different intracellular messengers. Interestingly, Bloomfield and Young (19) reported that Li enhanced the release of inflammatory mediators from PMN from patients with psoriasis compared to controls. PMN from such patients are reported to be "activated" and Li treatment sometimes exacerbates the disease. These authors postulate that the in vivo effect in patients may be explained by the in vitro findings of enhanced release of mediators. If the hypothesis is correct, then perhaps the effect of Li on PMN from patients with other autoimmune diseases (myasthesia gravis, rheumatoid arthritis, systemic lupus erythematosus, ankylosing spondylitis etc.) should be evaluated to determine if there is a consistent pattern of findings. Such data would be very informative since, as discussed earlier, Li treatment of patients with such diseases can lead to exacerbations of disease activity.

Relevant to this discussion are some observations in an animal model (mouse) of disease where enhanced release of PMN proteinases has been implicated (23). Treatment of mice with Li during the development and progression of PMN activation had no detectable effect on the in vivo plasma proteinase levels (24). Therefore, activation of PMN by different in vivo signals may not be consistently enhanced by Li.

Li has also been reported to depress zymosan-stimulated PMN chemiluminescence but the ion did not influence fMLP-mediated stimulation (25). Therefore, the effect of Li was again stimulus-dependent, indicating that Li was influencing a second messenger system rather than the indicator system (chemiluminescence). Using bacteria as a stimulant of the respiratory burst, we have found that 5 to 10 mM LiCl usually enhanced chemiluminescence by approximately 20% and 20 mM inhibited the response (Hart et al., unpublished). However, the enhancing response was variable and was not seen with PMN from some individuals. In this regard, our results were not unlike those of Siegal et al. (26) who studied zymosan and phorbol myristate acetate-activated PMN.

Natural Killer Cells

Natural killer (NK) cells are believed to be important for killing tumor cells as well as other immunoregulatory activities. These cells are bone marrow derived in mammals and have some characteristics in common with several other cells of bone marrow origin. Lithium lactate has been reported to have no effect on mouse NK activity in vitro (27). However, treatment of mice with Li lactate in vivo led to higher levels of NK activity in the spleens of such animals (27). Thus, Li does not appear to influence NK cells but it does appear to accelerate NK development from precursors. While Li may not influence NK activity directly, treatment of mice with lithium oxybutyrate prevented stress-induced depression of splenic NK activity (28).

Reports dealing with NK cell numbers or activity in patients receiving Li have not been found. However, it is likely that Li could exhibit effects on the integrity of the NK system in humans since it is known that the system can be influenced by mediators of stress reactions which modulate lithium-sensitive systems such as cAMP.

Mast Cells

Mast cells and IgE are central to host defense against parasites. Mast cells and IgE are also involved in unwanted immediate hypersensitivity reactivity and certain aspects of asthma. The intracellular second messenger systems associated with mast cell degranulation and activation are similar to those involved in PMN degranulation and activation (cAMP, inositol triphosphates, diacyglycerol, Ca^{++}-flux, G-proteins) and some of them have been postulated to be targets of Li action. However, a review of the literature has revealed studies which indicate that Li has no influence on asthma (29). Interestingly, reports in the literature of lithium allergy are very likely due to induction of sensitivity to nonlithium ingredients in commercial preparations (30). With regard to studies or reports dealing with mast cells specifically, there appears to be a paucity of information. Either studies dealing with Li effects on mast cells have not yielded interesting results or the failure of Li to have any influence on mast cell-dependent conditions (allergy, parasitic disease) in vivo has inhibited further in vitro studies.

Macrophages/Monocytes

Macrophages and monocytes play an important role in host defense. Their roles include phagocytosis of particles and microorganisms, secretion of growth factors and inflammatory mediators, and antigen processing and presentation to lymphocytes. In contrast to PMN, macrophages are a heterogenous set of cells (Kupfer cells, alveolar macrophages, splenic macrophages, bone-marrow macrophages, inflammatory macrophages, etc.) which can be induced to further differentiation of function by soluble mediators or bacterial products such as lipopolysaccharide.

Lithium has been reported to influence both the phagocytic activity of macrophages as well as the secretory activity of these cells. Shenkman et al. (31) have reported that LiCl increases the phagocytic activity of human peripheral blood monocyte/macrophages. As discussed earlier, the effect of Li on enhancement of granulopoiesis in bone-marrow cell cultures could be abrogated by removal of a phagocytic adherent cell type (16), presumably a macrophage. Li apparently enhances colony stimulating factor secretion from these cells. Addition of Li to cultures of hamster and rat plastic adherent bone marrow cells leads to enhanced secretion of neutral endoproteinases (Hart, unpublished). Whether these enzymes are from macrophages specifically is not yet known conclusively.

The effect of Li on a third important function of macrophages, namely antigen presentation/assisting lymphocyte activation, is not well characterized. Stimulation of lymphocytes by mitogens and antigens is dependent on macrophages for both growth factors and presentation of antigen in association with Class-II histocompatibility antigens (antigen-specific). Depletion of immune hamster lymph node populations of macrophages leads to a loss of antigen stimulation (26) and lithium could not restore responsiveness. Therefore, Li cannot replace macrophages under these conditions. Similar experiments carried out with macrophage-depleted human peripheral blood lymphocyte populations have led to the same conclusion (Matheson and Hart, unpublished observations). It is well known that expression of Class-II histocompatibility antigens by macrophages is a reversible process that can be modulated by a number of signals. However, it is not known whether Li can influence the expression of such antigens on macrophages or whether it can modify the antigen "processing" step.

Lymphocytes

Treatment of humans and experimental animals with Li salts has led to detectable changes in lymphoid tissues or lymphocyte activities. These latter findings include involution of the thymus of both normal and adrenalectomized mice (33), concentration of in vivo administered Li^+ in the cortex of the rat thymus (34), lymphopenia (10), and the appearance of antinuclear antibodies in a significant percentage of treated patients compared to age-matched controls (13). In the latter study, Li^+ treatment of a patient with an autoimmune disease, systemic lupus erythematosus (SLE), led to an exacerbation of the disease (13). In addition, two reports have appeared which indicate that Li_2CO_3 treatment of patients with hairy cell leukemia (B lymphocyte) can lead to a reversal of this disease (35, 36).

Human Peripheral Blood Mononuclear Cells

In 1978, Schenkman et al. (25) reported that addition of Li^+ to cultures of human peripheral blood lymphocytes enhanced the proliferative response of these cells to PHA. Concentrations of Li⁺ from 1.25 to 5 mM were found to be effective. In contrast to results obtained with the mitogen PHA, it was found that Li⁺ had less of an influence on the two-way mixed lymphocyte reaction (MLR). Only the highest concentration of Li⁺ tested, 5 mM, had an enhancing effect on the MLR. The response to another lectin, Con A, was also enhanced in the presence of Li⁺ (37). In other nonproliferation assays of lymphocytes, Schenkman et al. (31) also presented evidence that Li⁺ could enhance the percentage of sheep red blood rosette-forming cells (T-lymphocytes) found with peripheral blood. From experiments with compounds which elevate cAMP levels in cells, theophylline or prostaglandin PGE_2, it was concluded that the effect of Li⁺ was mediated by its ability to inhibit adenylate cyclase. Further results indicated that suppressor T-lymphocytes are the target of Li action (38).

Experiments similar to those just described have also been reported by Gelfand et al. (39). However, this latter group has pursued one aspect, the influence of Li⁺ on T-suppressor cells, in much greater detail (40, 41). With regard to PHA-induced mitogenesis, Gelfand et al. (33) reported that 5 mM Li⁺ could enhance ³H-thymidine incorporation by approximately 1.7-fold. In addition, 5 mM Li⁺ could reverse theophylline-induced suppression of the PHA response, indicating that Li⁺ was modifying the activity of adenylate cyclase. LiCl could also reverse the inhibition of IgM secretion induced by agents that elevate intracellular cAMP levels. Additional experiments by Gelfand et al. (40) and Dosch et al. (41), pertaining to the effect of Li⁺ on suppressor T-cell function, have led to the exciting possibility that Li⁺ could be used therapeutically to correct defects of humoral immunity mediated by this subpopulation of T-lymphocytes. LiCl was found to inhibit suppressor cell activity in vitro. Utilizing cells from immunodeficient patients with excess suppressor cell activity, these investigators reported that LiCl could functionally eliminate the influence of the suppressor cells (41). Prompted by their in vitro success, experiments were then performed in vivo with the same patients. Administration of Li⁺ led to an increase in the number of surface Ig positive B-lymphocytes, a decrease in the number of suppressor T-cells, but no increases in the serum concentration of IgM, IgE, or IgA. Therefore Li⁺ appears to only correct some of the characteristics associated with the humoral immunodeficiencies. Interestingly, Greco has reported that Li⁺ had little or no influence on immune reactivity in manic-depression patients or normals receiving lithium (42).

An additional set of variables regarding the effect of Li on human lymphoid cells was raised by the report of Licastro et al. (43). These workers reported that Li enhanced suboptimal mitogen stimulation of lymphocytes from older individuals or individuals with Down's syndrome (DS) to a much greater degree than cells from normal young adults. Since older individuals and DS subjects have reported alterations in immune regulation (suppressor cell activity), LiCl would appear to "correct" such alterations. However, the enhancement by LiCl was not completely specific for this ion, so the conclusion that Li was central to the effect cannot be made. Other reports in the literature (44, 45) also have indicated that stimulation of cells by suboptimal concentrations of stimuli was enhanced by Li. Fernandez and Fox (44) also concluded the effect was at the suppressor cell level. Bray et al. (45) reported that Li enhanced stimulation of peripheral blood lymphocytes by PHA and PWM and it was also effective in cultures that were made monocyte-deficient. Recently, in collaboration with Dr. David Matheson, the effect of Li on monocyte-depleted lymphocyte stimulation was pursued (Matheson and Hart, unpublished). Peripheral blood lymphocyte populations containing $15 \pm 5\%$, $2 \pm 0.5\%$, or 0.1% monocytes were stimulated with a range of PHA concentrations. The response in the presence of 0.1% monocytes was depressed ($> 90\%$) compared to the other cultures, again demonstrating the monocyte-dependence of this mitogen. LiCl ($2.5 - 10$ mM) did not replace the monocyte requirement and enhanced stimulation maximally 35%, 25%, and 10% in the presence of 15%, 2%, or 0.1% monocytes, respectively. These results would indicate that Li very likely does not enhance lymphocyte stimulation by stimulating the release of a rate-limiting growth factor from monocytes. Kucharz et al. (46) have reported that Li enhances IL-1 release from monocytes. If the primary effect of Li was at this level one might have expected a more dramatic enhancement under conditions where monocytes are limited.

Kucharz et al. (46) have also reported that Li enhances both IL-2 (a T-cell growth factor) release and lymphocyte responsiveness to the interleukin. The increased responsiveness was not apparently due to increased IL-2 receptor expression (46). Using normal human PBL (approximately 10% monocytes) it has been found that low concentrations of LiCl (2.5 mM), usually found to enhance mitogen stimulation, inhibited the response to pure human IL-2 (Matheson and Hart, unpublished). The basis for the difference between these results and those of Kucharz et al. (46) is unknown, but the latter group used an IL-2 dependent cell line, rather than normal cells, as an indicator system. Clarification of the mechanism(s) by which Li enhances stimulation of human lymphocytes must therefore await further investigation.

Lymphoid Cells from Laboratory Animals

Rats

Rats have been widely used to study the effects of lithium salts on behavior, brain metabolism, and neuroendocrine functions. Administration of Li to rats leads to accumulation in the thymus, particularly the cortex (34). Treatment of rats with lithium salts has been reported to influence, usually suppress, lymphocyte-dependent immune activities such as antibody formation and delayed type hypersensitivity (47,48). While in vivo experiments have indicated that Li can influence lymphocyte-dependent activities, there is not much literature available regarding effect of this ion on in vitro activity. Preliminary experiments carried out to investigate the effect of LiCl on mitogen stimulation of the rat splenocytes and lymph node cells have not yielded a reproducible pattern of influence (Hart, unpublished).

Guinea Pigs

Many of the early lithium experiments of John Cade were performed with guinea pigs (3). He found that lithium salts made the animals calmer and less responsive

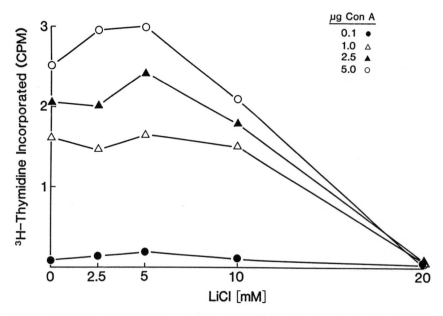

FIGURE 5.1. Effect of LiCl on Concanavalin A stimulation of guinea pig lymph node cells (LNC). LNC were stimulated for 72 hours with Con A in the absence or presence of 2.5, 5, 10, or 20 mM LiCl. Twenty-four hours prior to harvest the cultures were pulsed with ^{3}H-TdR. Triplicate cultures at each point were analyzed and variation was less than 10%.

to stimulation. While lithium salts have an influence on guinea pig behavior, the influence of this ion on immune activity in this experimental animal has not been very well investigated. Preliminary experiments from this laboratory have revealed that Li has a modest enhancing effect on Concanavalin A stimulation of guinea pig lymph node cells (LNC) when tested at low concentrations and an inhibitory effect at higher doses (Figure 5.1). Similar results have been obtained with splenocytes (Hart, unpublished). These effects were specific for Li and, interestingly, were not dependent on the concentration of mitogen used for stimulation. Unpublished results of experiments by Dr. R. Paque (University of Texas HSC at San Antonio) have revealed that Li$^+$ is capable of enhancing the in vitro responses of immune guinea pig LNC to soluble antigens (R. Paque, personal communication). However, the enhancement observed was extremely variable. Therefore, under the conditions employed for in vitro experiments, Li appears to exert modest, highly variable effects on lymphocyte stimulation, with the most consistent result being inhibition at high, nonphysiologic, concentrations.

Mice

Treatment of mice with lithium salts can lead to immunosuppression and immune deviation (49, 50). LiCl has been used in conditioned taste aversion experiments investigating links between such responses and immune reactivity (51). LiCl was found not to influence delayed type hypersensitivity reactions in nonconditioned mice (51). Experiments in vitro with lymphoid cells from BDF_1 mice have indicated that LiCl has no consistent, specific effect on lectin and lipopolysaccharide (LPS) stimulation of the cells (Hart, unpublished). Ishizaka and Moller (52) have reported that LPS stimulation of splenocytes from responsive mice (C3H/Tif) was not enhanced by LiCl, while LPS stimulation of cells from the LPS-resistant C3H/HeJ strain was enhanced by LiCl. The implication of this unconfirmed report is that Li can circumvent the genetic block in the cells of the C3H/HeJ strain but has minimal effects on normal lymphocyte responses. Interestingly, other investigators have reported that genetically different inbred strains of mice also vary in their sensitivity to the toxic effects of LiCl (53). As mice vary in their sensitivity to the toxic effects of the different isotopes of Li (^6Li and ^7Li) (54), perhaps the experiments dealing with the effect of Li on lymphocyte activity in mice should be reevaluated using pure ^7Li and a spectrum of genetically defined mouse strains.

Hamsters

For the past several years, the effect of Li^+ on the in vitro reactivity of hamster cells has been analyzed. This includes the proliferative response to both T- and B-lymphocyte mitogens, the proliferative response to antigens, and the development of antibody-forming cells. Early reports characterized the effect of Li^+ on the lymphoid cell response to PHA (55), Con A (56), Zn^{+2} (57), soluble protein antigens (32, 58), particulate antigens (59) and lipopolysaccharide (60, 61). Other reports have dealt with analyzing the mechanism by which Li^+ exerts an effect on these responses (11, 62, 63, 64). Because of the complexity of the findings I will attempt to present the results in a "stimulant" dependent fashion starting with the mitogens and ending with the antigen-driven responses.

In all tissues tested (thymus, lymph node, and spleen), LiCl enhanced the response of the cells to PHA (55). Optimal enhancement of ^3H-thymidine incorporation occurred at 10 mM LiCl. The effect was specific for Li^+ in that similar concentrations of K^+ or Na^+ were without effect. In addition, the effect also appeared to be independent of Ca^{+2}- or Mg^{+2}-dependent events. The influence of Li^+ on PHA stimulation was lost if addition was delayed until 20 hours poststimulation (55). LiCl was therefore influencing early events of stimulation. A few other points concerning the impact of Li^+ on PHA stimulation deserve some comment. The first is that Li^+ could enhance stimulation by suboptimal concentrations of PHA but could not induce stimulation by submitogenic concentrations of PHA (55). Therefore Li^+ may augment some rate-limiting process that is ancillary to the mitogenic signal but it cannot lower the threshold for PHA. Recently, experiments designed to determine the influence of adherent cells on PHA stimulation have revealed that making the cultures adherent-cell deficient by a single passage over Sephadex G-10 columns depresses the response to this mitogen (Table 5.1). The response of G-10 nonadherent cells to 25 µg or 50 µg of PHA

TABLE 5.1. Effect of Sephadex G-10 fractionation on the PHA responsiveness of hamster thymocytes.

Concentration of PHA (µg/ml)	Net CPM ^{3}H-TdR Incorporated			
	Unfractionated[a]		G-10 Nonadherent[b]	
	Control	10 mM Li$^+$	Control	10 mM Li$^+$
10	23520	85138 (362)[c]	12914	49185 (381)
25	41245	87381 (212)	14648	67000 (457)
50	40713	80195 (197)	13500	76853 (569)

[a]Incubated in medium + 5% FCS, washed 3X and cultured serum-free.
[b]5 × 10^8 cells in medium + 5% FCS fractionated on Sephadex G-10, washed 3X, cultured serum-free. Yield of cells from the Sephadex G-10 was 35–40%.
[c]Numbers in parentheses indicate % of control incorporation. ^{3}H-TdR incorporation into triplicate cultures varied less than 10%.

was depressed by 64% and 67%, respectively. Addition of 10 mM LiCl to the nonadherent cells led to 25 µg and 50 µg PHA responses that were 77% and 96% of the responses of the unfractionated cells + 10 mM LiCl. Therefore, under adherent cell-deficient conditions, the influence of Li$^+$ was more dramatic than under conditions where adherent cells were present in normal concentrations. While this area of experimentation is obviously not complete, the results thus far raise the possibility that Li$^+$ can influence stimulation via adherent cells. This finding is somewhat different from the results with human cells discussed earlier.

A second point concerning Li$^+$ and PHA stimulation deals with the biochemical mechanisms by which Li$^+$ influences stimulation of hamster cells. Evidence (although indirect) has been obtained that is consistent with Li$^+$ being able to modify both monovalent cation metabolism (56, 62, 63), as well as cyclic nucleotide metabolism (63). The former conclusion is based on the finding that LiCl enhancement of PHA stimulation in medium containing Rb$^+$ rather than K$^+$ is much lower than that observed in medium containing K$^+$ (62). Since Rb$^+$ can substitute for K$^+$ in many systems, it seems likely that Li$^+$ may be influencing K$^+$/Rb$^+$ metabolism possibly at the level of the NaK ATPase. The conclusion that Li$^+$ also can influence cyclic nucleotide metabolism is based on the findings that Li$^+$ could reverse inhibition increased by theophylline, dibutryrylcyclic AMP or indomethacin (63). These experiments, similar to those performed by Schenkman et al. (31) and Gelfand et al. (39), are again indirect but imply that Li$^+$ can inhibit hamster adenylate cyclase.

Characterization of the effect of Li$^+$ on stimulation induced by a second T-lymphocyte mitogen, Con A, yielded somewhat different results from those obtained with PHA (56, 63). First, Li$^+$ enhanced stimulation by suboptimal concentrations of Con A but accelerated the inhibition observed with supraoptimal concentrations of Con A (56). Similar to results with PHA, other effects of Li on Con A stimulation were: 1) specific for Li$^+$; 2) on early events of stimulation; 3) an enhancement of suboptimal stimulation but not induction of stimulation with submitogenic Con A concentrations. However, in contrast to PHA stimulation,

Con A stimulation was much more resistant to inhibition by cyclic AMP and agents that lead to increased cAMP concentrations (63). Two additional pieces of evidence have indicated that the effect of Li^+ could be partially reversed by increasing the extracellular K^+ concentration (56) and second, Li^+ enhanced ouabain inhibition of Con A stimulation while K^+ protected cells from ouabain inhibition (63). Therefore, with Con A as a stimulant, the primary system modified by Li^+ may very well be at the level of the NaK ATPase.

The two mitogens described are both plant lectins and both primarily T-lymphocyte mitogens. Investigation into the effect of Li^+ on lymphocyte activation has also been extended to include stimulation by B-lymphocyte mitogens. These investigations, using three mitogens, lipopolysaccharide, dextran sulfate, and trypsin, have yielded some interesting results. Probably the most interesting result is that of the three B-lymphocyte mitogens tested, only LPS stimulation was dramatically enhanced by Li^+ (60, 61). Dextran sulfate induced stimulation was not enhanced by Li^+ and stimulation by the protease trypsin was marginally enhanced. The Li^+-dependent enhancement of LPS stimulation had many of the same characteristics as had previously been observed for lectin stimulation. That is, 1) Li^+ appeared to modify early events of stimulation; 2) the effect was specific for Li^+ and was not observed with similar concentrations of NaCl, KCl, or RbCl; 3) maximum enhancement occurred at 10 mM Li^+; and 4) the Li^+ effect appeared to be mediated primarily at the level of cyclic nucleotide (cAMP) metabolism. With regard to the latter point, Li^+ could reverse theophylline, DBcAMP and indomethacin induced suppression of LPS stimulation (61). Experiments with ouabain were inconclusive (Hart, unpublished). Therefore the relationship of Li^+ enhancement of LPS stimulation to monovalent cation transport remains unclear.

The finding that stimulation of cells by dextran sulfate, and to a lesser extent trypsin stimulation, was resistant to Li^+ modulation is very intriguing and causes one to reconsider the possible site(s) of Li^+ involvement in stimulation. Several possibilities exist for these findings. The first is that dextran sulfate and trypsin modify the same systems influenced by Li^+ such that Li^+ can exert no additional effect. The second is that dextran sulfate and trypsin stimulate a subpopulation of cells that is relatively resistant to Li^+ influence. A third possibility is that the Li^+ effect is indirect; LPS, but not dextran sulfate or trypsin, may also modify the activity of an immunoregulatory cell which exerts an influence on lymphocyte stimulation. It is known that LPS can dramatically alter the activity of macrophages (65, 66). Activation of macrophages can lead to, among other things, production of prostaglandins which can inhibit lymphocyte stimulation (65, 66, 67). To investigate this possibility, low concentrations of indomethacin, known to inhibit prostaglandin synthesis, were added to cultures of splenocytes stimulated with LPS. There was some enhancement of LPS stimulation in the presence of low concentrations of indomethacin but it was much less than that observed with Li^+ (61). Therefore, it is unlikely that the effect of Li^+ on LPS stimulation can be attributed solely to prostaglandin production by macrophages.

Preliminary experiments utilizing cell separation techniques have recently indicated that Li^+ modifies the activity or influence of a subpopulation of cells that are separate from the majority of cells responding to LPS (Hart, unpublished). That is, a subpopulation of lymphocytes can be obtained which are responsive to LPS but resistant to Li$^+$ modulation, thereby supporting the hypothesis that Li$^+$ can also influence lymphocyte stimulation. One final point concerning Li$^+$ and LPS should be mentioned. Treatment of hamsters with 1 meq or 2.5 meq LiCl/kg body weight had no influence on LPS toxicity in vivo (Hart, unpublished). Hamsters are very resistant to LPS toxicity and, unlike mice, do not make a very efficient antibody response to LPS (68). LiCl treatment of hamsters prior to and during in vivo LPS challenge did not change the previous findings (Hart, unpublished). The dichotomy observed between in vivo and in vitro effects of Li$^+$ may be due to the presence of factors in vivo which overrides any potential effect (69).

While the studies with mitogens are interesting, they may not reflect the sequence of events accompanying antigen stimulation. Therefore, the effect of Li$^+$ on three assays of antigen-dependent activation was investigated. These include the two-way MLR, T-lymphocyte proliferation in response to a soluble protein antigen, and development of a primary in vitro response to a thymus-dependent antigen (sheep red blood cells). Addition of LiCl (1-25 mM) to mixed culture of lymph node cells from histoincompatible CB and MHA hamsters did not yield a consistent enhancement of ^{3}H-TdR incorporation (Hart, unpublished). In contrast to the MLR results, addition of 0.5 to 10 mM LiCl to antigen-stimulated cultures of normal hamster lymph node cells (70) led to an increase in the number of specific antibody-forming cells (AFC). This observed increase was antigen-dependent and the maximum increase (177% of control) occurred at 1 mM LiCl (59). The relative number of direct AFC detected with 0.5, 1.0, 2.5, 5, and 10 mM LiCl was 153%, 177%, 128%, 92%, and 21% of control, respectively. This dose response curve is somewhat different from that observed for mitogen stimulation where the maximum enhancement occurred at 10 mM LiCl. This difference may reflect a change between proliferation induced by mitogens and the activation and differentiation required for AFC development. It is obvious that Li$^+$ modulation of AFC development has not been completely characterized. Supplementary in vitro experiments delineating the effect of Li$^+$ on the response to thymus-independent antigens are also needed. In addition, the ability of Li$^+$ to influence humoral responses in vivo will be necessary in order to determine whether the in vitro findings can be extrapolated to the whole animal.

The third type of antigen-dependent stimulation that has been investigated is the in vitro proliferative response of immune T-lymphocytes to soluble protein antigens. Early experiments indicated that addition of LiCl to cultures of lymph node cells from animals immunized with either KLH or DNP-BSA led to potentiation of antigen-induced ^{3}H-TdR incorporation (58). Maximum potentiation occurred at 5 to 10 mM LiCl and was specific for Li$^+$. In addition, the Li$^+$ was only effective if added within the first 24 hours of the 72-hour culture (58). One disturbing aspect of Li$^+$ potentiation of antigen-dependent ^{3}H-TdR incorporation was the variation in the potentiation observed from experiment to experiment (58). The Li$^+$ dependent potentiation ranged from 30% to 90% in an early series of approximately 10 experiments. In a few later experiments little or no potentiation was observed with Li⁺, even when the cells were obviously immune. Such results seemed inconsistent with the concept that Li⁺ enhanced stimulation by a direct effect on antigen reactive lymphocytes and again raised the possibility that Li⁺ was influencing the activity or function of an accessory cell. This possibility was investigated using three approaches (32). First, lymph node cells from animals immunized with DNP-BSA in Freund's complete adjuvant (FCA) were fractionated on Sephadex G-10 to remove adherent cells. In contrast to previous results with PHA stimulation, removal of adherent cells by passage over G-10 completely abolished the in vitro response to antigen. The response to antigen could not be restored with either 5 or 10 mM LiCl. Therefore Li⁺ cannot replace the requirement for macrophages in antigen-dependent T-lymphocyte activation. The second approach utilized different immunization regimens. Animals were immunized in the footpads with antigen in saline, Freund's incomplete adjuvant (FICA), or FCA. Fourteen days postimmunization, cells from the draining lymph nodes were cultured with antigen ±5 to 10 mM LiCl. Lymph nodes from animals injected with antigen in saline were not enlarged and the cells did not respond to antigen even in the presence of LiCl. In contrast, immunization of animals with antigen in FICA did lead to the accumulation of antigen reactive cells but the proliferation was less than that observed with cells from animals immunized with antigen in FCA (32). When the cultures from animals immunized with antigen in FICA were supplemented with Li⁺, enhancement consistently occurred. In contrast, addition of Li⁺ to the cells from animals immunized with antigen in FCA had little or no effect on ³H-TdR incorporation. While such results could have several immunological interpretations, they do shed some insights into possible explanations for the variability seen in earlier experiments. Since commercially available adjuvant was utilized in these studies there may have been variability in the *Mycobacteria* content of the Freund's complete adjuvant. With regard to the immunological interpretation of the results, it would appear that Li⁺ can potentiate a suboptimal proliferative response but is relatively ineffectual when conditions are "optimal." Among the several possibilities for the cellular basis for the suboptimal conditions arising after FICA immunization, two seem very attractive. The first is that similar numbers of antigen-reactive cells are generated but optimal number of activated accessory cells (macrophages) which supply growth factors etc. are not generated. The second possibility is that excess suppressor cells are uniquely generated by the FICA regimen and Li⁺ negates their influence, similar to the system described by Gelfand et al. (40, 41) that was discussed previously. Additional experiments will be required to clarify this issue.

The third approach to characterizing Li⁺ enhancement of antigen-dependent proliferation evolved from the second approach described earlier. This involved attempts to correlate the degree of Li⁺ enhancement with time after immunization with antigen in FCA. The rationale is that there may be a differential decline in accessory cell function with time after immunization and a possible induction of suppressor cells. In a series of experiments, cells were cultured from animals immunized for 7 to 25 days. While maximum stimulation without Li⁺ occurred around 14 days postimmunization, Li+ appeared to exert a greater potentiating effect at the longer periods of time postimmunization (32). This again points to the conclusion that Li+ has a greater impact when conditions are not optimal.

One final point with regard to Li+ and the response to soluble antigen should be emphasized. That is, whatever the cellular basis for the Li+ effect, it appears to be related biochemically to cyclic nucleotide metabolism rather than other metabolic systems. This conclusion is based on the finding that addition of Li+ to cultures of antigen-reactive cells reversed both theophylline and DBcAMP induced inhibition of the proliferative response (32). Therefore, biochemically the Li+ effect on antigen-dependent stimulation is more closely related to PHA stimulation than Con A stimulation (63).

Conclusions Regarding the Effect of Li on Cells of the Immune System

Different Cell Types Vary Dramatically in Their Response to Lithium

Different elements of the immune system (PMN, NK cells, macrophages, lymphoid cells, etc.) respond to varying degrees in the presence of lithium. Such differences are very likely due to the indicator systems since such activities may or may not depend on lithium-dependent second messenger systems for performance of the indicator function. Thus, NK cell binding to a target cell and subsequent killing of the target is not dependent on lithium-sensitive systems while the events leading to phagocytosis of particles by macrophages do include lithium-sensitive steps. While Li is known to influence a number of second messenger systems (cAMP, G-proteins, ITP), such systems may or may not be critical for many cell functions or may not be in subcellular compartments involved in such functions.

Cell Types Vary in Their In Vitro and In Vivo Responses to Lithium

In several instances, cells of the immune system can be influenced by Li under in vitro conditions but in vivo there is little evidence that Li has the same impact. The possible explanations for this phenomenon could comprise a long list. However the most obvious ones include 1) removal of normal in vivo regulators which override the impact of Li; 2) use of Li concentrations in vitro that are supraphysiologic and which cannot be attained in vivo; 3) the in vitro system is static while in vivo the immune system is more dynamic; and 4) culture conditions in vitro are very suboptimal compared to in vivo conditions so that the effect of Li is magnified out of proportion. All of these factors probably impinge on our interpretation of results.

Lymphoid Cells from Different Species Vary in Their Responsiveness to Lithium's Influence

In the human, Li has been shown to influence abnormal regulation in vivo but the ion apparently has minimal impact under normal conditions. Thus, Li can only assist in the reestablishment of homeostasis (64). Reports from this laboratory, as well as a number of other laboratories, have found lymphocytes from species such as mouse, rat, and guinea pig to generally be unresponsive to Li in vitro or inhibited by the ion. In contrast, stimulation of lymphoid cells from hamsters was found to be consistently enhanced, in most circumstances, by supplementing the cultures with Li. The enhancement varied somewhat with the indicator system, but was found with cells from a number of lymphoid tissues (thymus, lymph node, and spleen) so was not localized to a specific environment. From these results one would have to conclude that in vitro Li cannot override mechanisms regulating lymphocytes from most species in spite of the fact that Li-modifiable second messenger systems (cAMP, Ca^{+2}, ITP, etc.) have been implicated in the activation process.

Possible Explanations for the Results Obtained with Hamster Lymphoid Cells

As discussed previously, stimulation of lymphocytes from hamsters was consistently enhanced by supplementing the cultures with 1 to 10 mM Li. For the most part, the degree of enhancement and the range of stimuli enhanced separates the hamster system from results obtained with cells from humans, mice, rats, and guinea pigs. Except for the results obtained with antigen-specific LNC derived from optimal in vivo activation, the experiments were performed with normal lymphocytes. Interestingly, cells derived from optimally stimulated animals were the least responsive to Li, while "resting" cells were the most responsive to the ion. Thus, in the maximally activated state, the cells from this species respond to Li in a fashion not unlike resting cells from the other species mentioned. Therefore the "resting" or maintenance state of the cells from hamsters is apparently unique.

Why the hamster cells should be different from the cells of other species could be just another example of normal species diversity, each with their own pattern of regulatory mechanisms developed for unknown reasons. Perhaps if lymphocytes from a large number of species were tested, others would be found which behave like hamster cells, others like human cells and others completely different from the two mentioned.

A second possibility for the unique responsiveness of hamster lymphoid cells to Li may actually relate to other aspects of biological regulation in hamsters. Of the species investigated thus far, the hamster is the only animal in the group that hibernates. Laboratory Syrian hamsters can be induced to exhibit a hibernating state by altering the photoperiod and the room temperature. However, hamsters are "optional" hibernators and are not strictly seasonal hibernators (71).

Hibernation in hamsters is accompanied by a depression in splenic immune responsiveness (72). Spleen fragments or splenocyte suspensions were unable to mount an antibody response to a thymus-dependent antigen, sheep red blood cells, or to elicit a graft versus host reaction in vitro, respectively (72). Whether this loss of activity is lymphocyte-related or related to accessory cells (Class II MHC antigen expression?) has not been determined. The spleen and thymus of the animals were not markedly morphologically different from tissues obtained from normal or cold-adapted animals. The tissues from hamsters were also not markedly involuted. In contrast, Shivatcheva and Hadjioloff (73) reported that gut-associated lymphoid tissue of the European ground squirrel undergoes a seasonal involution which is based partly on the circannual cycle and partly on hibernation. Results reported by Sidkey et al. (74) using the Richardson ground squirrel have also indicated that seasonal variation in immune responsiveness has two components. The first is a circannual depression and the second is an exaggeration due to the hibernation phase. In ground squirrels the immunodepression accompanying hibernation is not complete, but is severe (75, 76). It should be pointed out that many species exhibit circannual modulation of immune responsiveness but only a unique subset of species can/will proceed to the hibernation phase. Therefore cells from such species may have evolved unique regulatory mechanisms which separate them from other species. Such mechanisms could involve a unique dependence on normal stimuli, sensitivity to "hibernation"-related factors, or a loss of normal basal activity in order to accommodate the wide fluctuations in responsiveness required to maintain metabolic activity in the hibernating state. Hibernation is usually accompanied by profound hypothermia, respiratory depression, analgesia, and a general lowering of metabolic activity. Therefore, cells of the immune system of hibernating species would have to accommodate the lowered state and survival until aroused in order to not compromise the integrity of the animal. If such accommodation involved lithium-dependent steps then one would expect Li to influence hibernation. A survey of the literature revealed only one reference to the effect of Li on hibernation in hamsters (77). Unfortunately, in this study the concentration of Li used resulted in a 50% mortality in the animals and therefore the overt toxicity abrogated the development of any conclusions. Such a study should be repeated with other concentrations of Li since, as Zvolskey et al. point out (77), there are some interesting parallels between hibernation and primary affective disorders.

There are also interesting aspects of hibernation that relate to possible neuroendocrine-immunoregulation interactions that are relevant to the effect of Li on immune cells. In recent years the interfacing of the immune system with the neuroendocrine system has been the subject of many studies (78–81). This has led to exciting developments in the fields of neuroimmunology and psychoneuroimmunology. This is relevant to the discussion since a number of the mechanisms possible to be influential in circannual and hibernation relate to this interface and may be a "common ground" for Li action on regulatory mechanisms, particularly with regard to why the hamster system appears to be different from the others investigated thus far.

The mechanisms underlying circannual phenomenon and the subsequent hibernation process are not completely understood. However, it is obvious that the pineal gland is intimately involved. Products of the pineal gland, such as melatonin, have been shown to influence circadian immune activity (82, 83) and very likely this gland also influences seasonal variations in biologic activity. Products of the pituitary and hypothalamus are also probably involved in regulation as well. Interestingly, some strains of mice such as C57B1/6, whose cells undergo seasonal changes (84), are apparently unable to make melatonin in their pineal gland (84). Thus other factors (or sources of melatonin-like molecules) must also be involved. Other workers have postulated that brown fat could contribute to neuroimmunomodulation through, as yet, unknown mechanisms (86, 87). How such tissue mediates immunomodulation is unclear, especially at the level of hibernators versus nonhibernators. Some animals in hibernation do have a circulating plasma factor which can induce hibernation in summer active mammals (88). This plasma component has been termed "hibernation induction trigger" but its source and mechanism of action has not been well defined. While biochemically not well characterized, such triggers are not species specific since HIT in plasma from hibernating bears can induce hibernation in ground squirrels (88). A direct effect of such triggers on elements of the immune system has not been investigated, but this possibility should be explored. Interestingly, Syrian hamsters do not apparently elaborate a plasma HIT (71).

A final explanation (at least for this review) for species differences is related to innervation of lymphoid tissues and "organs." Lymphoid tissues such as thymus, spleen and lymph nodes, bone marrow, and gut-associated lymphoid tissue are innervated (89–97 and references therein). Disruption of the integrity of this interface by procedures such as chemical sympathectomy can lead to alterations in immune responsiveness and regulation (90, 93–96). Interestingly, Felton et al. (96) have reported some strain-dependent differences in response to chemical sympathectomy. Chemical sympathectomy of BALB/c mice led to a diminished T-dependent response while in C57BL/6 mice no diminution was noted. Therefore, a genetic component, other than pineal gland melatonin synthesis (85), is implicated and no simple generalizations can be made. Interestingly, lithium has been reported to lengthen free-running circadian rhythms in both of these strains (98), so many variables are at work.

Reports of innervation studies of the lymphoid tissues of hibernating animals were not found, nor were reports of chemical sympathectomy effects on hibernating animals identified. However, Korneeva and Serdiukova (99) have reported that seasonal changes occur in the levels of catacholamines in nerves and other tissues of the hamster. It would be of interest to determine whether this phenomenon also occurs in lymphoid tissues of the hamster and other hibernating mammals. All of the studies on hamster lymphoid cell responses and lithium have used innervated tissues (thymus, spleen, and lymph node cells). While experiments using cells from innervated tissues of other species have not revealed dramatic Li effects, most of the studies with human lymphocytes have used peripheral blood cells which are obviously in a noninnervated compartment.

From the previous discussion, it is obvious that animals that hibernate, or have the potential to hibernate, must have evolved unique mechanisms to regulate a number of systems, including the immune system, without compromising species survival. Such modifications appear to be superimposed on circannual regulatory systems. Both sets of regulatory mechanisms are not well understood mechanistically but it is apparent that nonhibernators and hibernators alike exhibit seasonal variations, but lymphoid cells from species such as humans and mice do not exhibit dramatic responses to Li in vitro or in vivo unless there is an alteration in normal regulatory function (10–12, 17, 19, 64). Therefore, one could speculate that species adaptation to hibernation has led to alterations in regulatory systems such that they may be viewed as abnormal when compared to human and mouse "prototypes."

Future Direction

It is readily apparent from recent developments in the fields of neuroimmunoendocrinology and psychoneuroimmunology that there are interrelationships and parallels between the systems we have been discussing. Lithium may very well interact at this interface. The degree of activity at this interface may be related to a number of variables, some of which have been addressed or alluded to during the previous discussion. While part of this discussion was admittedly speculative, the concepts raised can be tested and modified. For instance, does Li influence cells of the immune system from hibernating hamsters as well as nonhibernating animals? Do lymphoid cells from other hibernating species respond to Li as well as hamsters or can one not generalize from the observations on hamster cells? What is the effect of nontoxic concentrations of Li on initiation and maintenance of hibernation? Can we learn new insights into lithium's effect on behavior and affective disorders by studying how this ion influences the complex interactions of immune cells? If indeed there are both parallels and interrelationships between the neuro- and immuno-systems, then insights into the variations possible may provide clues to better identify subpopulations of patients as well as insights into new therapeutic directions. Further identification of lithium-sensitive steps on the neuro-side of the coin may have potential application for the control of immunodysregulation such as chronic inflammatory diseases or in the tumor-bearing state.

Acknowledgments. The author thanks Judy Crawford for her excellent secretarial assistance in the preparation of the manuscript. Some of the investigations discussed were supported by the Canadian Arthritis Society, the Alberta Cancer Board, the American Cancer Society, and the Alberta Heritage Foundation for Medical Research. The author is supported by the AHFMR.

References

1. Spite, M., & Spite, F. (1982). Lithium abundance at the formation of the galaxy. *Nature (Lond.), 297,*483–485.
1a. Bach, R.O., Ellestad, R.B., Kamienski, C.W., & Wasson, B.P. (1981). Lithium and lithium compounds. In Kirk-Othmer *Encyclopedia of chemical technology,* Vol. 14, (3rd Ed.) (pp. 448–476). New York: John Wiley and Sons.

2. Johnson, F.N. (1984). *The history of lithium therapy.* London: Macmillan.
3. Johnson, F.N. (1985). The early history of lithium therapy. In R.O. Bach (Ed.), *Lithium: Current applications in science, medicine and technology* (pp. 337–344). New York: John Wiley and Sons.
4. Cade, J.F. (1949). Lithium salts in the treatment of psychotic excitement. *Med. J. Australia, 36,*349–353.
5. Crammer, J.L. (1985). The problem of lithium mechanisms. In S. Gaby, J. Harris, & B. Ho (Eds.), *Metal ions in neurology and psychiatry* (pp. 165–176). New York: Alan R. Liss.
6. Birch, N.J. (Ed.). (1988). Lithium: Inorganic pharmacology and psychiatric use. Oxford: IRL Press.
7. Hodgkinson, S., Sherrington, R., Gurling, H., Marchbanks, R., Reeders, S., Mallet, J., McInnis, M., Petursson, H., & Brynjolfsson, J. (1987). Molecular genetic evidence for heterogeneity in manic depression. *Nature (Lond.), 325,*805–806.
8. Baron, M., Risch, N., Hamberger, R., Mandel, B., Kushner, S., Newman, M., Drumer, D., & Belmaker, R. (1987). Genetic linkage between X-chromosome markers and bipolar affective disorders. *Nature (Lond.), 326,*289–292.
9. Smith, D.F. (1977). *Lithium and animal behavior.* Montreal: Eden Press.
10. Rossof, A., & Robinson, W. (Eds.) (1980). Lithium effects on granulopoiesis and immune function. New York: Plenum Press.
11. Hart, D.A. (1986). Lithium as an *in vitro* modulator of immune cell function: Clues to its *in vivo* biological activities. *IRCS Medical Science, 14,*756–762.
12. Lieb, J. (1987). Lithium and immune function. *Med. Hypotheses, 23,*73–93.
13. Presley, A., Kahn, A., & Williamson, N. (1976). Antinuclear antibodies in patients on lithium carbonate. *Brit. Med. J., 2(6080),*280–281.
14. Levitt, L., & Quesenberry, P. (1980). The effect of lithium on murine hematopoiesis in liquid culture. *N. Engl. J. Med., 302,*713–717.
15. Gallicchio, V., & Chen, M. (1981). Influence of lithium on proliferation of hematopoietic stem cells. *Exp. Hematol., 9,*804–810.
16. Richman, C., Kinnealey, A., & Hofman, P. (1981). Granulopoietic effects of lithium on human bone marrow *in vitro. Exp. Hematol., 9,*449–455.
17. Perez, H.D., Kaplan, H., Goldstein, I., Shenkiman, L., & Borkowsky, W. (1980). Reversal of an abnormality of polymorphonuclear leukocyte chemotaxis with lithium. *Clin., Immunol. Immunopathol., 16,*308–315.
18. Bloomfield, F., & Young, M. (1982). Influence of lithium and fluoride on degranulation from human neutrophils *in vitro. Inflammation, 6,*257–267.
19. Bloomfield, F., & Young, M. (1983). Enhanced Release of Inflammatory Mediators from Lithium-Stimulated Neutrophils in Psoriasis. *Brit. J. Dermatol., 109,*9–13.
20. Hart, D.A., Groenewoud, Y., Rabin, H., & Chamberland, S. (1986). Lithium induced enzyme release from polymorphonuclear leukocytes derived from rodents, normal humans or patients with cystic fibrosis is isotope independent. *IRCS Med. Sci., 14,*213–214.
21. Hart, D.A., Groenewoud, Y., & Chamberland, S. (1986). Characterization of lithium-induced enzyme release from human polymorphonuclear leukocytes. *Biochem. Cell Biol., 64,*880–885.
22. Murphy, P., & Hart, D.A. (1987). Regulation of enzyme release from human polymorphonuclear leukocytes: Further evidence for the independent regulation of granule subpopulations. *Biochem. Cell. Biol., 65,*1007–1015.
23. Hart, D.A. (1988). Differences between beige and bg/+ mice in the disruption of plasma proteinase regulation in the tumor-bearing state following *C. parvum* treat-

ment: Evidence for the involvement of polymorphonuclear leukocyte proteinases. *Haemostasis, 18,*154–162.

24. Hart, D.A. (1987). Steroids and tuftsin fail to prevent the induction of altered plasma proteinase homeostasis in mice bearing the B16 melanoma or treated with *C. parvum. Int. J. Immunopharmacol., 9,*669–674.

25. Rossini, D., Vaglini, F., Petrini, M., & Bertelli, A. (1987). Lithium versus rubidium in chemiluminescence. *Acta Haemato., 77,*241–242.

26. Siegal, J., Johnston, R., Lowe, R., Epstein, P., & Rossof, A. (1980). Effects of lithium on neutrophil metabolism *in vitro* and on neutrophil function during therapy. In A. Rossof & W. Robinson (Eds.), *Lithium effects on granulopoiesis and immune function* (pp. 371–388). New York: Plenum Press.

27. Fuggetta, M., Alvino, E., Romani, L., Grohman, U., Potenza, C., & Giuliani, A. (1988). Increase of natural killer activity of mouse lymphocytes following *in vitro* and *in vivo* treatment with lithium. *Immunopharmacol. Immunotoxicol., 10,*79–91.

28. Kryzhanoskii, G., & Sukhikh, G. (1986). Prophylactic effects of sodium oxybutyrate and lithium oxybutyrate on stress-induced depression of the activity of normal killers in mice (translated title). *Biull. Ekop. Biol. Med. (Russian), 102,*592–593.

29. Spitz, E., Saltz, H., & Bearman, J. (1982). A double blind cross-over trial of lithium carbonate in asthma. *Ann. Allergy, 49,*165–168.

30. Clark, K., & Efferson, J. (1987). Lithium allergy. *J. Clin. Psychopharmacol., 7,*287–289.

31. Shenkman, L., Borkowsky, W., Holazman, R., & Shopsin, B. (1978). Enhancement of lymphocyte and macrophage function *in vitro* by lithium chloride. *Clin. Immunol. Immunopathol., 10,*187–192.

32. Hart, D.A. (1988). Lithium potentiates antigen-dependent stimulation of lymphocytes only under suboptimal conditions. *Int. J. Immunopharmac., 10,*153–160.

33. Perez-Cruet, J., & Dancy, J.T. (1977). Thymus gland involution induced by lithium chloride. *Experientia, 33,*646–648.

34. Nelson, S., Herman, M., Bensch, K., Shu, R., & Barchas, J. (1976). Localization and quantitation of lithium in rat tissue following intraperitoneal injection of lithium chloride. *Exptl. Mol. Path., 25,*38–48.

35. Blum, S.F. (1980). Lithium and hairy cell leukemia. *N. Engl. J. Med., 303,*464–465.

36. Paladine, W., Price, L., Williams, H., & Jevtic, M. (1981). Hairy cell leukemia treated with lithium. *N. Engl. J. Med., 304,*1237–1238.

37. Shenkman, L., Borkowsky, W., & Shopsin, B. (1980). Lithium as an immunologic adjuvant. *Med. Hypotheses, 6,*1–6.

38. Wadler, S., Shenkman, L., & Borkowsky, W. (1979). Effects of lithium on suppressor-enriched and suppressor-depleted mononuclear cell preparations. *Clin. Res., 27,*339.

39. Gelfand, E., Dosch, H-M., Hastings, D., & Shore, A. (1979). Lithium: A modulator of cyclic AMP-dependent events in lymphocytes? *Science, 203,*365–367.

40. Gelfand, E., Cheung, R., Hastings, D., & Dosch, H-M. (1980). Characterization of lithium effects on two aspects of T-cell function. In A. Rossof & W. Robinson (Eds.), *Lithium effects on granulopoiesis and immune function* (pp. 429–446). New York: Plenum Press.

41. Dosch, H-M., Matheson, D., Shuurman, R., & Gelfand, E. (1980). Anti-suppressor cell effects of lithium *in vitro* and *in vivo*. In A. Rossof & W. Robinson (Eds.), *Lithium effects on granulopoiesis and immune function* (pp. 447–462). New York: Plenum Press.

42. Greco, F.A. (1980). Lithium and immune function in man. In A. Rossof & W. Robinson (Eds.), *Effects on granulopoiesis and immune function* (pp. 463–469). New York: Plenum Press.

43. Licastro, F., Chiricolo, M., Tabacchi, P., Barboni, F., Zannotti, M., & Franceschi, C. (1983). Enhancing effect of lithium and potassium ions on lectin-induced lymphocyte proliferation in aging and Down's syndrome subjects. *Cell. Immunol., 75*,111–121.
44. Bray, J., Turner, A.R., & Dusel, F. (1981). Lithium and the mitogenic response of human lymphocytes. *Clin. Immunol. Immunopathol., 19*,284–288.
45. Fernandez, L., & Fox, R. (1980). Perturbation of the human immune system by lithium. *Clin. Exp. Immunol., 41*,527–532.
46. Kucharz, E., Sierakowski, S., Staite, N., & Goodwin, J. (1988). Mechanism of lithium-induced augmentation of T-cell proliferation. *Int. J. Immunopharma., 10*,253–259.
47. Jankovic, B.D., Lenert, P., & Mitrovic, K. (1982). Suppression of experimental allergic thyroiditis in rats treated with lithium chloride. *Immunobiology, 161*,488–493.
48. Jankovic, B.D., Popeskovic, L., & Isakovic, K. (1978). Cation-induced immunosuppression: The effect of lithium on arthus reactivity, delayed hypersensitivity and antibody production in the rat. *Adv. Exptl. Biol. Med., 114*,339–344.
49. Jankovic, B.D., Popeskovic, L., & Isakovic, K. (1978). Suppressed immune responses in mouse and rat during treatment with lithium chloride. *Fed. Proc., 37*,1651.
50. Popeskovic, L., Jankovic, B.D., & Isakovic, K. (1979). Effect of lithium on immunological reactivity in CBA mice. *Period. Biol., 81*,179.
51. Kelley, K., Dantzer, R., Mormede, P., Salmon, H., & Aynaud, J. (1985). Conditioned taste aversion suppresses induction of delayed-type hypersensitivity immune reactions. *Physiol. Behav., 34*,189–193.
52. Ishizaka, S., & Moller, G. (1982). Lithium chloride induces partial responsiveness to LPS in non-responder B cells. *Nature, 299*,363–365.
53. El-Kassem, M., & Singh, S. (1983). Strain dependent rate of Li^+ elimination associated with toxic effects of lethal doses of lithium chloride in mice. *Pharmacol. Biochem. Behav., 19*,257–261.
54. Alexander, G., Lieberman, K., & Stokes, P. (1980). Differential lethality of lithium isotopes in mice. *Biol. Psychiatry, 15*,469–471.
55. Hart, D.A. (1979). Potentiation of phytohemagglutinin stimulation of lymphoid cells by lithium. *Exptl. Cell. Res., 119*,47–53.
56. Hart, D.A. (1979). Modulation of concanavalin A stimulation of hamster lymphoid cells by lithium chloride. *Cell. Immunol., 43*,113–122.
57. Hart, D.A. (1979). Augmentation of zinc ion stimulation of lymphoid cells by calcium and lithium. *Exptl. Cell. Res., 121*,419–425.
58. Hart, D.A. (1979). Modulation of lymphocyte activation by LiCl. In J.G. Kaplan (Ed.), *The molecular basis of immune cell function* (pp. 408–411). Amsterdam: North-Holland.
59. Hart, D.A., & Stein-Streilein, J. (1981). Hamster lymphoid responses *in vitro*. In J.W. Streilein, D.A. Hart, J. Stein-Streilein, W. Duncan & R.E. Billingham (Eds.), *Hamster immune responses in infectious and oncologic diseases* (pp. 7–22). New York: Plenum Press.
60. Hart, D.A. (1982). Differential potentiation of lipopolysaccharide stimulation of lymphoid cells by lithium. *Cell. Immunol., 71*,159–168.
61. Hart, D.A. (1982). Studies on the mechanism of LiCl enhancement of lipopolysaccharide stimulation of lymphoid cells. *Cell. Immunol., 71*,169–182.
62. Hart, D.A. (1981). Ability of monovalent cations to replace potassium during stimulation of hamster lymphoid cells. *Cell. Immunol., 57*,209–218.
63. Hart, D.A. (1981). Evidence that lithium ions can modulate lectin stimulation by multiple mechanisms. *Cell. Immunol., 58*,372–384.

64. Hart, D.A. (1988). Immunopharmacologic aspects of lithium: One aspect of a general role as a modulator of homeostasis. In N.J. Birch (Ed.), *Lithium: Inorganic Pharmacology and Psychiatric Use* (pp. 99–102). Oxford: IRL Press.

65. Vogel, S., Marshall, S., & Rosenstreich, D. (1979). Analysis of the effects of lipopolysaccharide on macrophages: Differential phagocytic responses of C3H/HeN and C3H/HeJ macrophages *in vitro*. *Infect. Immun.*, *25*,328–336.

66. Hart, D.A. (1988). Age and 1pr dependent induction of increased sensitivity to the toxic effects of lipopolysaccharide and indomethacin in MRL mice. *J. Lab. Clin. Immunol.*, *26*,129–134.

67. Goodrum, K., Moore, R., & Berry, L. (1978). Effect of indomethacin on the response of mice to endotoxin. *J. Reticuloendothel Soc.*, *23*,213–221.

68. Streilein, J., & Hart, D.A. (1976). Serum-free culture of hamster lymphoid cells and differential inhibition of LPS stimulation by isologous serum. *Infect. Immun.*, *14*, 463–470.

69. Streilein, J., & Hart, D.A. (1978). The role of alpha globulins in non-specific regulation of the immune response: Possible mechanism for external and internal signals. *Fed. Proc.*, *37*,2042–2044.

70. Stein-Streilein, J., & Hart, D.A. (1979). *In vitro* development of a primary antibody response with dissociated cells from hamsters, guinea pigs and mice: Evidence that the cells responsible reside primarily with lymph node cells. *Cell. Immunol.*, *45*, 241–248.

71. Minor, J., Bishop, D., & Badger, C. (1978). The golden hamster and the blood-borne hibernation trigger. *Cryobiology, 15*,557–562.

72. Sidky, Y., & Auerbach, R. (1968). Effect of hibernation on the hamster spleen immune response reaction *in vitro*. *Proc. Soc. Exptl. Biol. Med.*, *129*,122–127.

73. Shivatchea, T., & Hadjioloff, A. (1987). Seasonal involution of gut-associated lymphoid tissues of the european ground squirrel. *Dev. Comp. Immunol.*, *11*,791–799.

74. Sidky, Y., Hayward, J., & Ruth, R. (1972). Seasonal variations of the immune response of ground squirrels kept at 22–24°C. *Canad. J. of Physiol. Pharmacol.*, *50*, 203–206.

75. McKenna, J., & Musacchia, X. (1968). Antibody formation in hibernating ground squirrels. *Proc. Soc. Exptl. Biol. Med.*, *129*,720–724.

76. Jaroslow, B. (1968). Development of the secondary hemolysin response in hibernating ground squirrels (Citellus Tridecemlineatus). *Proc. Natl. Acad. Sci.*, *61*,69–76.

77. Zvolsky, P., Jansky, L., Vyskocilova, J., & Grof, P. (1981). Effects of psychotropic drugs on hamster hibernation-pilot study. *Prog. Neuro-Psychopharmacol.*, *5*,599–602.

78. Ader, R. (1981). *Psychoneuroimmunology*. New York: Academic Press.

79. Blalock, E., & Smith, E. (1985). The immune system: Our mobile brain? *Immunol. Today, 6*,115–117.

80. Blalock, J., & Bost, K. (Eds.). (1988). *Neuroimmunoendocrinology*. Basel: Karger.

81. Jankovic, B., Markovic, B., & Spector, N. (Eds.). (1987). Neuroimmune interactions. *Ann. N.Y. Acad. Sci.*, Vol. 496.

82. Maestroni, G., Conti, A., & Pierpaoli, W. (1987). The role of the pineal gland in immunity. *Clin. Exp. Immunol.*, *68*,384–391.

83. Pierpaoli, W., & Maestroni, G. (1987). Melatonin: A principal neuroimmunoregulatory and anti-stress hormone. *Immunol. Lett.*, *16*,355–361.

84. Brock, M. (1987). Seasonal changes in recovery of cryopreserved murine lymphocytes resemble endogenous rhythms of unfrozen cells. *Cryobiology, 24*,412–419.

85. Ebihara, S., Marko, T., Hudson, D., & Menaker, M. (1986). Genetic control of melatonin synthesis in the pineal gland of the mouse. *Science, 231*,491–493.
86. Sidky, Y., Daggett, L., & Auerbach, R. (1969). Brown fat: Its possible role in immunosuppression during hibernation. *Proc. Soc. Exptl. Biol. Med., 132*,760–763.
87. Jankovic, B. (1987). Brown adipose tissue: Its *in vivo* immunology and involvement in neuroimmunomodulation. *Ann. NY. Acad. Sci., 496*,3–26.
88. Oeltgen, P., Welborn, J., Nuchols, P., Spurrier, W., Bruce D., & Su, T-P. (1987). Opioids and hibernation: Effects of Kappa Opioid U69593 on induction of hibernation in summer-active ground squirrels by "hibernation induction trigger" (HIT). *Life Sci., 41*,2115–2120.
89. Calvo, W. (1968). The innervation of the bone marrow in laboratory animals. *Amer. J. Anat., 123*,315–328.
90. Kasahara, K., Tanaka, S., Ito, T., & Hamashima, Y. (1977). Suppression of the primary immune response by chemical sympathectomy. *Res. Commun. Chem. Path. Pharm., 16*,687–694.
91. Giron, L.T., Crutcher, K., & Davis, J. (1980). Lymph nodes—A possible site for sympathetic neuronal regulation of immune responses. *Ann. Neurol., 8*,520–525.
92. Bulloch, K., & Moore, R.Y. (1981). Innervation of the thymus gland by brain stem and spinal cord in mouse and rat. *Amer. J. Anat., 162*,157–166.
93. Williams, J., & Felton, D. (1981). Sympathetic innervation of murine thymus and spleen: A comparative histofluorescence study. *Anat. Rec., 199*,531–542.
94. Williams, J., Peterson, R., Shea, P., Schmetje, J., Bauer, D., & Felton, D. (1981). Sympathetic innervation of murine thymus and spleen: Evidence for a functional link between the nervous and immune systems. *Brain Res. Bull., 6*,83–94.
95. Hall, N., McClure, J., Hu, S-K., Tare, N., Seals, C., & Goldstein, A.L. (1982). Effects of 6-hydroxydopamine upon primary and secondary thymus dependent immune responses. *Immunopharmacol., 5*,39–48.
96. Felton, D., Livnat, S., Felton, S., Carlson, S., Bellinger, D., & Yeh, P. (1984). Sympathetic innervation of lymph nodes of mice. *Brain Res. Bull., 13*,693–699.
97. Singh, U. (1984). Sympathetic innervation of fetal mouse thymus. *Eur. J. Immunol., 14*,757–759.
98. Possidents, B., & Exner, R. (1986). Gene-dependent effect of lithium on circadian rhythms in mice *(Mus musculus). Chronobiol. Int., 3*,17–21.
99. Korneeva, T.E., & Serdiukova, E. (1987). Seasonal changes in the adrenergic fibers in the wall of the microvessels of the cheek pouch of the hamster, *Mesocricetus auratus* (translated title). *Zh. Evol. Biokhim. Fiziol., 23*,642–646.

6
Lithium and Granulopoiesis: Mechanism of Action

VINCENT S. GALLICCHIO

Introduction

Lithium has been used clinically for more than a century and for the past four decades has been the treatment of choice for certain forms of manic and depressive illnesses (1,2). Although lithium is the lightest metal known to exist in nature, it commonly is not considered to be one of the trace biological elements for mammalian cell systems. (This topic is discussed in detail in Chapter 1). However, it is a well-documented observation that lithium is capable of producing effects on the hematopoietic system, particularly on the production of granulocytes (granulopoiesis), which was first observed in manic depressive patients receiving lithium as therapy (3). Although the topic of lithium and hematopoiesis has been reviewed previously (4–6) it has been more than five years since the last major in-depth review was conducted. Much more information has been generated on the subject of lithium and hematopoiesis, primarily focusing on the production of granulocytes, and the use of lithium in conditions where the production of granulocytes is either faulty or inadequate.

Effects of Lithium on Blood Cells

The production of granulocytes of leukocytosis, although documented since the early 1970's, has only been thoroughly investigated within the last decade. In summary of the many studies conducted to identify the lithium concentration required to induce leukocytosis that investigated both normal subjects as well as psychiatric patients on lithium therapy, it is concluded that a blood plasma or serum lithium level of 0.3 mEq/L or greater is sufficient to produce and sustain a peripheral blood leukocytosis (7–10). This leukocytosis in specific cellular composition has included increased numbers of neutrophils, eosinophils, and to a lesser degree monocytes. Cells not influenced in overall numbers are basophils. Lithium appears to be effective in decreasing the concentration of lymphocytes causing a lymphopenia (lithium effects on lymphocytes are discussed elsewhere in this volume, see Chapter 5). An interesting observation is the reported ability of lithium to increase the circulating platelet level both from animal studies reported by our laboratory (11) as well as from human subjects (7). The effect of lithium of circulating platelets has not been fully explored in clinical conditions where platelet production is inadequate and therefore is an area of continued investigation. The ability of lithium to induce changes in the circulating red cell count appears to be universal: lithium has only a partial inhibitory response on blood erythrocyte and reticulocyte levels (12,13).

Lithium Stimulation of Neutrophil Production

The documented increase in neutrophil production usually does not increase by any factor greater than 1.5 times baseline for any given individual. Following lithium administration in humans, neutrophil production increases steadily for the first week and then reaches a plateau. Most accurate studies to determine the precise lithium concentration required to achieve maximum neutrophil production have not identified the optimum lithium concentration that needs to be reached in order to stimulate granulocyte production in vivo. Lithium carbonate ingestion ranging from 600 to 1200 mg/day has produced significant increases in total neutrophil populations correlation with a serum lithium level in the range from 0.2 to 0.9 mEq/L (14). Neutrophilia has been demonstrated to develop following lithium administration in a wide variety of animal species, besides humans, that includes mice, rats, hamsters, and dogs (4,5).

The kinetics of neutrophil production have been examined in a study that involved 12 patients treated with a lithium carbonate dose of 900 to 1800 mg/day (9). The total blood granulocyte pool was increased by a factor of almost 2 over controls. These observations were critical because they demonstrated that the lithium-induced neutrophilia was not just simply due to an increased left shift of cells from the invascular marginating pool of cells into the circulation but were due to an expansion of the proliferating cells of the bone marrow.

Effect of Lithium on Bone Marrow Cell Populations

The observations that the lithium stimulation of neutrophilia involved a true proliferative response rather than just a shift of cell populations from the marginating to the circulatory pool of cells, lead investigators to begin to examine the bone marrow for changes in rates of mitotic cell proliferation (15,16). These studies concluded that the rise in neutrophil populations following lithium were due in part to the increased production of the mitotic and postmitotic pool of neutrophil precursors. This expansion of neutrophil precursors identified that the population of cells responsible for increased neutrophil production involved the bone marrow progenitor stem cell pool (see Fig 6.1).

Evidence accumulated rapidly from animal studies (mice and dogs) and humans that identified lithium as capable of increasing the number of the granulocyte

MECHANISM OF LITHIUM ACTION

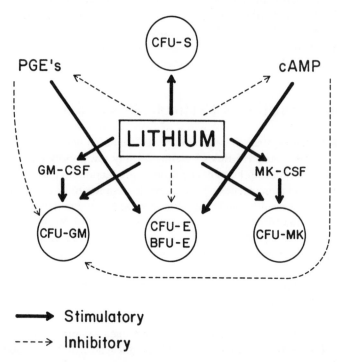

— Stimulatory

----→ Inhibitory

FIGURE 6.1. The effects and potential routes of lithium action on hematopoietic stem cells. Key: CFU-S, multipotential stem cell; CFU-GM, granulocyte-macrophage progenitor; CFU-E BFU-E, erythroid progenitors; CFU-MK, megakaryocyte progenitor; GM-CSF, granulocyte-macrophage colony stimulating factor; MK-CSF, megakaryocyte colony stimulating factor; PGE, prostaglandin; cAMP, cyclic nucleotides.

precursor progenitor stem cell, the colony forming unit granulocyte-macrophage (CFU-GM), located in the bone marrow, that is directly responsible for the production of more differentiated granulocyte progeny (17–21). This effect of lithium on CFU-GM was identified to involve the mitotic pool of these progenitor cells because their proliferation in vitro was stimulated following lithium exposure as measured by suicide studies using thymidine and hydroxyurea (22).

The ability of lithium to increase CFU-GM was shown by various investigators to involve lithium stimulation of CFU-GM dependency upon the presence of the growth factor required for CFU-GM survival known as colony stimulating factor (GM-CSF). In fact, lithium effectively increased the production of colony stimulating factor in a variety of different tissues (17,23,24). Studies from our laboratory (24) demonstrated that the ability of lithium to increase GM-CSF production from both mitogen-induced spleen and thymus cells prepared as serum-free conditioned mediae was both a mitogen and a specific lithium mediated event. Identical cultures prepared with added Na, K, Ca, or Mg did not induce GM-CSF activity when compared to those prepared with lithium. These studies demonstrated the unique capacity of lithium to influence the differentiation of committed hematopoietic stem cells, possibly by modulating the production of such growth factors required for hematopoietic differentiation.

However, the ability of lithium to increase neutrophil production by stimulating the proliferation of CFU-GM and the increased production of GM-CSF would not necessarily prove to be the only mechanism whereby lithium influences blood cell production. Because progenitor stem cells are present in bone marrow that have the CFU-GM as their progeny, it could be argued that the ability of lithium to increase the pool of granulopoietic progenitors like CFU-GM could be due to the ability of lithium to act on the multipotential stem cell known as the CFU-S, to increase its proliferation and thus would ultimately produce more cells capable of differentiation. It is a well-agreed fact that the multipotential stem cell, CFU-S, is the cell that is responsible for all of the differentiated lineage progeny that constitutes the hematopoietic system. In fact, studies performed by our laboratory demonstrated that lithium was capable of increasing the total number of CFU-S harvested from bone marrow when lithium was either administered in vivo in mice (12) or when murine bone marrow cells were incubated with lithium in vitro (18). Additional in vitro studies performed using the long-term marrow culture system demonstrated that lithium, when added to these cultures, increased the long-term survival and number of CFU-S harvested (25).

As has been described lithium can augment various aspects of granulopoiesis. Because lithium is a monovalent cation it is likely that its action on granulopoietic cells may involve transport processes that are a function of lithium permeability (for a review of transport mechanisms of the lithium ion refer to Chapter 4 in this volume). Studies conducted in our laboratory (26) identified the importance of sodium transport pathways in the mechanism whereby lithium increases CFU-GM, incorporating the use of agents known to alter sodium ion permeability. At noncytotoxic concentrations and after various timed exposures, gramicidin (an Na specific ionophore) and valinomycin (a K specific ionophore) were added to bone marrow cell cultures before the addition of lithium. In the presence of either ionophore the ability of lithium to increase CFU-GM was reduced when compared to lithium control cultures. However, what was significant was the observation that in the cultures plated in the presence of gramicidin the number of CFU-GM was less reduced than in the presence of valinomycin.

To further examine the role of sodium permeability in the mechanism whereby lithium stimulates in vitro granulopoiesis, additional studies were performed using more specific Na transport inhibitors such as amiloride and phloretin. Amiloride is an effective inhibitor of the passive sodium pump in a number of tissues and is used as a K sparing diuretic. Phloretin is an effective agent for inhibiting Na ion permeability. In the presence of either agent the ability of lithium to stimulate CFU-GM was significantly inhibited. These results demonstrated a role for sodium transport pathways as a possible mechanism for lithium augmentation on in vitro granulopoiesis and assumedly in vivo as well, since only in the presence of sodium transport inhibitors was there a significant reduction of the effect of lithium to stimulate CFU-GM.

Although sodium transport can also influence intracellular Ca by elevating Ca release from intracellular storage pools, thus implying Ca may be involved in lithium-induced granulopoiesis, bone marrow cultures plated in the presence of the antibiotic Ca ionophore A23187 produced a significant reduction in CFU-GM that was not reversed in the presence of added lithium (26). In fact in the presence of lithium added Ca inonphore A23187 inhibited lithium increased CFU-GM thus indicating Ca may be involved in the ability of lithium to promote granulopoiesis in vitro. This is an especially intriguing observation since in the presence of added Ca, events related to red cell production are enhanced whereas these same reactions are inhibited by lithium. Therefore concentrations of various monovalent and divalent cations existing in the marrow milieu may direct, if not significantly influence, what specific cell lineage progenitor cells differentiate.

The fact that the marrow milieu or microenvironment may be involved in the ability of lithium to influence granulopoiesis has been investigated (11). The marrow microenvironment or "stroma" has been studied to determine whether the addition of lithium altered the stroma of the bone marrow, spleen, or both. Lithium produces a significant increase in the number of both marrow and splenic derived stromal colonies following in vivo administration. Also these stromal colonies were effective in their capacity to support the in vitro growth of CFU-GM and platelet progenitors CFU-Meg (colony forming unit— megakaryocyte). These studies clearly demonstrate the ability of lithium to influence the microenvironment of both the bone marrow and the spleen by its capacity to increase stromal cell elements in steady-state conditions and indicate a role for the involvement of an active stroma which can be enhanced by lithium, and indicate lithium use may be of value in hematopoietic conditions that are influenced by an inadequate or faulty microenvironment.

Additional studies that have investigated the role of lithium in modulating granulocyte production incorporating the use of the long-term marrow culture system have identified lithium influences on the same target cells that were implied from the in vivo experiments. In these parallel studies (27,28) exposure of lithium (4 mEq/L) induced the production of factors elaborated from the adherent stroma cell layer which were capable of supporting the growth of granulocytes, macrophages, megakaryocytes, and mixed cells. Furthermore, this growth factor production which increased in the presence of lithium was blocked by an antibody to both GM-CSF and CSF1. CSF1 is a growth factor specific for the growth of macrophages rather than granulocytes. From these investigations it appears lithium acts upon an adherent marrow stroma cell layer to produce myeloid regulatory growth factors that are effective in supporting the maintenance of myeloid progenitor cells in vitro.

Cyclic nucleotides have received attention in the overall role they may play in the mechanism that regulates hematopoiesis in general, but also with respect to the action of lithium since it is an effective inhibitor of adenylcyclase activity.

Lithium, by its action of stimulating granulopoiesis, also inhibits erythropoiesis as has been previously indicated. When add in vitro lithium effectively inhibits erythroid progenitor cell proliferation at the same concentrations that are effective in stimulating in vitro granulopoiesis (29). The inhibition of adenylcyclase activity by lithium essentially restricts the formation of dibutyryl cyclic AMP (cdAMP), which either by itself or with other known agents capable of increasing its concentration, is a known stimulator of erythropoiesis. Conversely, dcAMP or any agonist are effective inhibitors of granulopoiesis, therefore lithium by inactivating adenylcyclase would promote reactions responsible for granulopoiesis rather than erythropoiesis. Thus lithium stimulation of granulopoiesis is being accomplished by the decreased intracellular concentrations of dcAMP. A study designed to investigate this hypothesis (30) demonstrated that the addition of various adenine nucleotides which stimulated granulopoiesis in vitro, increased it significantly in the presence of lithium; however, the addition of dcAMP only modestly reduced granulopoiesis and was ineffective in blocking lithium stimulated in vitro granulopoiesis. These studies indicate the complex interactions that are involved between monovalent cations, nucleotides, and the control of mammalian cell proliferation.

The Role of Lithium in Conditions Where Granulopoiesis Is Either Faulty or Inadequate

The overall implications from many of these investigations were that lithium can influence numerous hematopoietic events leading to an increased production of granulocytes; therefore, the use of lithium may be warranted as an effective adjuvant in many situations where neutropenia is a clinical problem. A useful model that indicated lithium may have clinical utility is the grey collie dog which inherits cyclic hematopoiesis as a recessive characteristic. All myeloid type cells are afflicted such that these animals become severely neutropenic for various cycles within their lifetime. The animals become susceptible to increasing infections and bleeding episodes ultimately leading to death. The use of lithium in the model was effective in restoring all circulating blood cells to near normal (31). Lithium use in the comparable human condition has not produced the same overall uniform results as observed with the dog model (20).

Based on the observed effects lithium compounds have on various aspects of hematopoiesis, numerous studies have been conducted where lithium was given to patients with aplastic anemia. Although with only marginal success in cases of aplastic anemia, lithium can be effective in neutropenic conditions where the actual etiological defect can be attributed to either an inadequate level of GM-CSF or where the myelosuppression is due to an increasingly active suppressor cell lymphocytre population (32,33).

A number of hematological conditions associated with faulty hematopoiesis resulting in neutropenic conditions of various etiologies where lithium treatment has proven successful have been Felty's syndrome (34); cyclic neutropenia in humans (35), and congenital and acquired neutropenia (36–38). It appears throughout many of the clinical studies performed that for lithium to be effective in neutropenic patients depends on the presence of a residual stem cell population capable of responding to lithium. Lithium therapy will fail if these patients have very few or no granulopoietic precursor stem cells to serve as the appropriate target cell population. If this condition exists lithium is ineffective in its ability to reverse any neutropenic condition.

It has been apparent that the most promising clinical use for lithium, with respect to neutropenic conditions, is supported by the results of clinical trials and animal studies demonstrating the effectiveness of lithium to reverse the hematopoietic toxicity associated with the therapy administered for malignant disease, whether it be chemotherapy or radiation (36–51). Investigations performed at our laboratory (11,52) have demonstrated two important observations in the mechanism of lithium-stimulated recovery of granulopoiesis following cytotoxic radiation administration: 1) Lithium-stimulated recovery of granulopoiesis following sublethal irradiation is a stem cell response and not mediated via increased levels of GM-CSF and 2) Lithium mediates its stimulation of granulopoietic recovery by accelerating the reconstitution and functional status of the hematopoietic inductive microenvironment or stroma.

In summary, these studies demonstrate that when lithium is administered prior to cytotoxic drugs and or radiation, lithium can reduce the degree of both neutropenia and thrombocytopenia apparently by limiting the degree of drug-induced stem cell toxicity. Furthermore, these studies stress the importance of the pretreatment period, since during this time modulating stem cells and their progeny may induce the necessary reactions that allow these stem cells to escape the toxicities commonly associated with the use of anticancer drug and radiation therapy. These studies further indicate a role of lithium use as a therapeutic agent in conditions where neutrophil and/or platelet production is either inadequate or faulty.

Most recent results from our laboratory have identified another hematological area for potential lithium use that is related to its role in manic depressive illness. Although lithium is used in the treatment of affective disorders to reduce the frequency and severity of manic depressive illness, certain drawbacks are known to exist, such as its low therapeutic index and the tendency to induce side effects in some patients. For these reasons, alternative substances have been investigated for their ability to provide the same degree of therapeutic effectiveness with less toxicity. Recent studies have indicated that carbamazepine (CBZ), an imipramine-like anticonvulsant, may also be an extremely effective agent in the treatment of a variety of psychiatric conditions. Although CBZ has gained a growing respectability as an extremely effective treatment for a wide variety of psychiatric conditions, it unfortunately can produce side effects such as leukopenia and aplastic anemia (53). We have identified that the leukopenia associated with CBZ treatment is due to the ability of CBZ to inhibit bone marrow progenitor cells (CFU-GM and CFU-Meg) and demonstrated that lithium can reverse the marrow toxic effects of CBZ (54). It appears that combined lithium-CBZ may provide synergistic psychotropic effects at the same time that lithium is capable of minimizing the potential of CBZ-induced marrow toxicity.

Effect of Lithium on the Induction of Leukemia

Because lithium has been associated with the production of granulocytes, concerns have been raised over the past several years as to the ability of lithium to stimulate unrestricted granulocyte production which could lead to the induction of leukemia. This concern has been raised as the result of observations where lithium has been used as adjuvant treatment in retarding the myelosuppression associated with the use of chemotherapy or radiation in cancer therapy (54). Patients described in many of these various studies were afflicted with either acute or chronic leukemia. Treatment with lithium in these patients demonstrated a recurrence of disease as measured by an increase in the percentage of leukemic blast cells present in peripheral blood. However, many other studies have been reported using lithium as adjuvant therapy for myelosuppression in leukemic patients without any reappearance or rapid development of disease (55). This question—whether sustained in vivo lithium administration would result in the development of leukemia—has been investigated in our laboratory (56). The studies were designed to investigate lithium effects on granulopoiesis in vivo by examining the effect of continuous lithium administration in a murine model. Lithium produced a significant elevation in the WBC at all time points examined (weeks 3–10). The degree of stimulation varied from 156% to 456% of control. However, by week 12 WBC values from the lithium-administered group were within control values. The total amount of lithium consumed by each mouse ranged from a low of 0.82 ± 0.02 mM (week 3) to 1.05 ± 0.03 mM (week 10). The volume intake produced a serum lithium level that ranged between 1.06 ± 0.5 mM to 1.21 ± 0.4 mM. There were no adverse signs in animal health nor any abnormal hematological changes on peripheral blood differential analysis. Bone marrow stem cells were significantly increased during the course of the examination, however by week 10 levels had returned to baseline. These studies demonstrated the capacity of lithium to sustain granulopoiesis in vivo over a period of up to 10 weeks without any adverse signs of toxicity or mortality. The serum lithium level was within the therapeutic range throughout the study period. We conclude that with close monitoring to avoid any potential toxicity-induced problems, long-term lithium use in situations where faulty or impaired granulopoiesis is a problem may be successfully achieved and maintained.

Although there have been numerous reports on the success of lithium use in retarding the myelosuppression of cancer therapy (55), a report worth mentioning is indicative of the positive effects lithium may have in this area. Lithium carbonate was used in 24 patients with acute leukemia (19 cases of acute myeloid leukemias, 3 cases of acute promyelocytic leukemia, and 2 cases of acute lymphoid leukemia) to minimize the degree and extent of chemotherapy-induced granulocytopenia and therefore reduce the risk of infection. In all patients the degree and duration of the neutropenic period was reduced in these lithium-treated patients along with a significant reduction of febrile episodes. However, this response of lithium has unfortunately not been universally observed in every clinical trial (56) and remains one of the limiting factors for greater acceptance and wider use of lithium in all leukemic conditions.

Areas of Future Promise

Two areas that lithium use may hold future promise are 1) use in AIDS patients to reverse the faulty lymphocyte production associated with AIDS (57) and the ability to help combat the burden of increased febrile episodes by augmenting the production of interleukin-2 and neutrophil activation, and 2) use in bone marrow transplantation to effectively increase the rate of bone marrow engraftment when administered to the donor before marrow harvest to minimize the degree and extent of suppressed hematopoiesis. This has been successfully accomplished using an animal model system for syngenic transplantation (58).

In summary, granulocyte production can be influenced by the administration of lithium salts. Lithium effectively enhances granulopoiesis both in vivo and in vitro and at the same concentrations stimulates platelet production and inhibits erythropoiesis. Many of the clinical trials with lithium demonstrate that lithium is effective in disorders that are characterized by an inadequate production of GM-CSF and when there is still a population of marrow stem cell progenitors to effectively serve as the target cell population. Recent studies have demonstrated that an important component of the ability of lithium to influence blood cells involves the transport of cations across the cell membrane and suggest such processes play an important role in normal cell proliferation and differentiation.

References

1. Fawcett, J. (1980). Lithium carbonate in medicine and psychiatry. In A.H. Rossof & W.A. Robinson (Eds.), Lithium effects on granulopoiesis and immune function. (p. 1) New York: Plenum Press.
2. Schou, M. (1957). Biology and pharmacology of the lithium ion. *Pharmacol. Rev., 9*, 17.
3. Shopsin, B., Friedman, R., & Gershon, S. (1971). Lithium and leukocytosis. *Clin. Pharmacol. Ther., 12*, 923.
4. Boggs, D.A., & Joyce, R.A. (1983). The hematopoietic effects of lithium. *Sem. Hematol., 20*, 129.
5. Barr, R.D., & Galbraith, P.R. (1983). *Can. Med. Assoc. J., 128*, 123.
6. Gallicchio, V.S. (1988). Lithium stimulation of granulopoiesis: Mechanism of action. In N.J. Birch (Ed.), *Lithium: Inorganic pharmacology and psychiatric use* (p. 93). Oxford: IRL Press.
7. Bille, P.E., Jensen, M.K, Jensen, J.P.K., & Poulsen, J.C. (1975). Studies on the hematologic and cytogenetic effects of lithium. *Acta Med. Scand., 198*, 281.
8. Perez-Cruet, J., Dancey, J.T., & Warte, J. (1977). Lithium effects on leukocytosis and

lymphopenia. In F.N. Johnson (Ed.), *Lithium in medical practice*, 1st British Lithium Congress, Univ. of Lancaster, UK.

9. Rothstein, G., Clarkson, D.R., Larsen, W., Grasser, B.L., & Athens, J.W. (1978). Effect of lithium on neutrophil mass and production.

10. Stein, R.S., Howard, C.A., & Brennar, M. (1981). Lithium carbonate and granulocyte production: Dose optimization. *Cancer, 48,* 2696.

11. Gallicchio, V.S., Gamba-Vitalo, C., Watts, T.D., & Chen, M.G. (1986). In vivo and in vitro modulation of megakaryocytopoiesis and stromal colony formulation by lithium. *J. Lab. Clin. Med., 108,* 199.

12. Gallicchio, V.S., & Chen, M.G. (1980). Modulation of murine pluripotential stem cell proliferation in vivo by lithium carbonate.

13. Chan, H.S.L., Saunders, E.F., & Freedman, M.H. (1980). Modulation of human hematopoiesis by prostaglandins and lithium. *J. Lab. Clin. Med., 95,* 125.

14. Ricci, P., Bandini, G., & Franchi, P. (1981). Haematological effects of lithium carbonate: A study of 56 psychiatric patients. *Haematologica, 66,* 62.

15. Malloy, E.L., Zauber, N.P., & Chervenick, P.A. (1978). The effect of lithium on blood and marrow neutrophils. *Blood* (Abstract), 52, 1280.

16. Joyce, R.A., & Chervenick, P.A. (1980). Lithium effects on granulopoiesis in mice following chemotherapy. In A.H. Rossof & W.A. Robinson (Eds.), *Lithium effects on granulopoiesis and immune function* (p. 145). New York: Plenum Press.

17. Harker, W.G., Rothstein, G., Clarkson, D., Athens, J.W., & MacFarlane, J.L. Enhancement of colony-stimulating activity production by lithium. *Blood, 49,* 263.

18. Gallicchio, V.S., & Chen, M.G. (1981). Influence of lithium on proliferation of hematopoietic stem cells. *Exper. Hematol., 9,* 804.

19. Rossof, A.H., & Fehir, K.M. (1979). Lithium carbonate increases marrow granulocyte committed colony forming units and peripheral blood granulocytes in a canine model. *Exper. Hematol., 7,* 255.

20. Hammond, W.P., & Dale, D.C. (1982). Cyclic hematopoiesis: Effects of lithium on colony-forming cells and colony-stimulating activity in grey collie dogs. *Blood, 59,* 179.

21. Morley, D.C., & Galbraith, P.R., (1978). Effect of lithium on granulopoiesis in culture. *Can. Med. Assoc. J., 118,* 288.

22. Gallicchio, V.S., & Chen, M.G. (1982). Cell kinetics of lithium-induced granulopoiesis. *Cell Tiss. Kinet., 15,* 179.

23. Ramsey, R., & Hays, E.F. (1979). Factors promoting colony stimulating activity (CSA) production in macrophages and epithelioid cells. *Exp. Hematol., 7,* 245.

24. Gallicchio, V.S., Chen, M.G., & Watts, T.D. (1984). Specificity of lithium (Li+) to enhance the production of colony stimulating factor (GM-CSF) from mitogen-stimulated lymphocytes in vitro. *Cellul. Immunol., 85,* 58.

25. Levitt, L.G., & Quesenberry, P.J. (1980). The effect of lithium on murine hematopoiesis in a liquid culture system. *N. Engl. J. Med., 302,* 713.

26. Gallicchio, V.S. (1986). Lithium stimulation of in vitro granulopoiesis: Evidence for mediation via sodium transport pathways. *Brit. J. Haematol., 62,* 455.

27. McGrath, H.E., Liang, C.M., Alberico, T.A., & Quesenberry, P.J. (1987). The effect of lithium on growth factor production in long-term bone marrow cultures. *Blood, 70,* 1136.

28. Quesenberry, P.J., Song, Z., Alberico, T., Gualtieri, R., Stewart, M., Innes, D., McGrath, H.E., Cranston, S., & Kleeman, E. (1985). Bone marrow adherent cell hemopoietic growth factor production. *Prog. Clin. Biol. Res., 184,* 247.

29. Gallicchio, V.S., & Murphy, M.J., Jr. (1983). Cation influences on in vitro growth of erythroid stem cells (CFU-E and BFU-E). *Cell Tissue Reser., 233*, 175.
30. Gualtieri, R.J., Berne, R.M., McGrath, H.E., Huster, W.J., & Quesenberry, P.J. (1986). Effect of adenive nucleotides on granulopoiesis and lithium-induced granulocytosis in long-term bone marrow cultures. *Exp. Hematol., 14*, 689.
31. Hammond, W.P., & Dale, D.C. (1980). Lithium therapy of canine cyclic hematopoiesis. *Blood, 55*, 26.
32. Barret, A.J., Hugh-Jones, K., Newton, K., & Watson, J.G. (1977). Lithium therapy in aplastic anemia. *Lancet, 1*, 202.
33. Blum, S.F. (1980). Lithium therapy of aplastic anemia. *N. Engl. J. Med. 300*, 677.
34. Gupta, R.C., Robinson, W.A., & Kurnick, J.E. (1976). Felty's syndrome—Effect of lithium on granulopoiesis. *Amer. J. Med., 61*, 29.
35. Verma, D.S., Spitzer, G., Zander, A.R., Dickie, K.A., & McCredie, K.B. (1972). Cyclic neutropenia and T lymphocyte suppression of granulopoiesis: Abrogation of the neutropenic cycles of lithium carbonate. *Leuk. Res., 6*, 567.
36. Gupta, R.C., Robinson, W.A., & Smyth, C.G. (1975). Efficacy of lithium in rheumatoid arthritis with granulocytopenia (Felty's Syndrome). *Arthritis Rheum., 18*, 179.
37. Barrett, A.J. (1980). Clinical experience with lithium in aplastic anemia and congenital neutropenia. *Adv. Exp. Biol. Med., 127*, 305.
38. Yassa, R., & Ananth, J. (1981). Treatment of neuroleptic-induced leukopenia with lithium carbonate. *Can. J. Psychiat., 26*, 487.
39. Joyce, R.A., & Chervenick, P.A. (1980). Lithium effects on granulopoiesis in mice following cytotoxic chemotherapy. *Adv. Exp. Biol. Med., 127*, 145.
40. Greco, R.C., Robinson, W.A., & Kurnick, J.E. (1976). Effect of lithium on granulopoiesis. *Amer. J. Med., 61*, 29.
41. Catane, R., Kaufman, J., Mittleman, A., & Murphy, G.P. (1977). Attenuation of myelosuppression with lithium. *N. Engl. J. Med., 297*, 452.
42. Lyman, G.H., Williams, C.G., & Preston, D. (1980). The effect of lithium carbonate administration in patients with advanced small cell bronchogenic carcinoma receiving combination chemotherapy and radiotherapy. *Adv. Exp. Biol. Med., 127*, 207.
43. Steinhertz, P.G., Rosen, G., Ghavimi, F., Wang, Y., & Miller, D.R. (1980). The effect of lithium carbonate on leukopenia after chemotherapy. *N. Engl. J. Med., 302*, 257.
44. Cass, C., Turner, A.R., Steiner, M., Analunis, M.J., & Tarr, T. (1981). Effect of lithium on the myelosuppression and chemotherapeutic activities of vinblastine. *Cancer Res., 41*, 1000.
45. Chan, H.S.L., Freedman, H., & Saunders, E.F. (1980). Stimulation of granulopoiesis in vitro and in vivo using lithium in children with chronic neutropenia. *Adv. Exp. Biol. Med., 127*, 293.
46. Gallicchio, V.S., Chen, M.G., & Watts, T.D. (1984). Ability of lithium to accelerate the recovery of granulopoiesis after subacute radiation injury. *Acta Rad. Oncol., 23*, 361.
47. Gallicchio, V.S. (1986). Lithium and hematopoietic toxicity I. Recovery in vivo of murine hematopoietic stem cells (CFU-S and CFU-Mix) after single-dose administration of cyclophosphamide. *Exper. Hematol., 14*, 395.
48. Gallicchio, V.S. (1987). Lithium and hematopoietic toxicity II. Acceleration in vivo of murine hematopoietic progenitor cells (CFU-GM and CFU-Meg) following treatment with vinblastine sulfate. *Int. J. Cell Clon., 5*, 122.
49. Gallicchio, V.S. (1988). Lithium and hematopoietic toxicity III. In vivo recovery of hematopoiesis following single-dose administration of cyclophosphamide. *Acta Haematol., 79*, 192.

50. Vacek, A., Sikulova, J., & Bartonickova, A. (1982). Radiation resistance in mice increased following chronic application of Li_2CO_3. *Acta Rad. Oncol., 21,* 325.
51. Stein, R.S., & Howard, C.A. (1980). Lithium therapy of chronic neutropenia. *Adv. Exp. Med. Biol.,* Vol. 127. New York: Plenum Press.
52. Gallicchio, V.S., Chen, M.G., & Watts, T.D. (1985). Lithium-stimulated recovery of granulopoiesis after sublethal irradiation is not mediated via increased levels of colony stimulating factor (CSF). *Int. J. Rad. Biol., 47,* 581.
53. Pisciotta, A.V. (1975). Hematological toxicity of carbamazepine. In J.K. Penry & D.D. Daly (Eds.), *Advances in neurology,* Vol. II (p. 355). New York: Raven Press.
54. Gallicchio, V.S., & Hulette, B.C. (1989). In vitro effect of lithium on carbamezapine-induced inhibition of murine and human bone marrow derived granulocyte macrophage, erythroid, and megakaryocyte progenitor stem cells. *Proc. Soc. Exp. Biol. Med., 190,* in press.
55. Rossof, A.H., & Robinson, W.A. (1980). Lithium effects on granulopoiesis and immune function. *Adv. Exp. Med. Biol.,* Vol. 127. New York: Plenum Press.
56. Gallicchio, V.S., & Watts, T.D. (1985). Sustain elevation of murine granulopoiesis in vivo with lithium carbonate. *IRCS Med. Sci., 13,* 1051.
57. Stein, M.B., Simon, G.L., Parenti, D.M., Scheib, R., Goldstein, A.L., Goodman, R., DiGioia, R. Paxton, H., Skotnicki, A.B., & Schulof, R.S., (1987). In vitro effects of thymosin and lithium on lymphoproliferative responses of normal donors and HIV seropositive male homosexuals with AIDS-related complex. *Clin. Immunol. Immunopathol., 44,* 51.
58. Messino, M.J., Gallicchio, V.S., Hulette, B.C., Friedman, D., Beischke, M., Gass, C., & Doukas, M.A. The effect of lithium to enhance engraftment of bone marrow progentors in a syngeneic murine transplant model. *Exper. Hematol.,* in press.

7
Does the Effect of Lithium on G-Proteins Have Behavioral Correlates?

ROBERT H. BELMAKER, SOFIA SCHREIBER-AVISSAR,
GABRIEL SCHREIBER, ZEV KAPLAN, YORAM GIVANT,
PESACH LICHTENBERG, AND JOSEPH ZOHAR

Lithium is a unique treatment in psychiatry with both antidepressant and anti-manic properties. The mode of action of lithium has not yet been clarified. Lithium has been shown to inhibit the accumulation of cyclic AMP stimulated by noradrenaline both in intact rat brain synaptosomes and in slices from rat cortex after in vitro or ex vivo administration (1,2). This effect occurs within the therapeutic range of the drug (0.7–1.5 mM) and has been proposed to account for its clinical efficacy (3). Hormone-sensitive adenylate cyclase activity is the result of the concerted function of a complex of at least three separate proteins: a hormone receptor, a catalyst, and a guanine nucleotide-binding protein (4). It is well established that beta-adrenergic-induced activation of adenylate cyclase is mediated by the stimulatory G-protein, Gs. Upon agonist binding, a quaternary complex is formed between agonist-occupied receptor and GTP-bound G-protein. Subsequently, the G-protein is dissociated into subunits (5). The alpha-s subunit activates adenylate cyclase while exhibiting GTP-ase activity. Several authors investigated lithium action on the various components along the hormone-activated adenylate cyclase pathway. Although lithium inhibits agonist-stimulated adenylate cyclase activity, it has been shown not to affect beta-adrenergic receptor density (6). Lithium was shown to inhibit Gpp (NH) p-stimulated adenylate cyclase activity in rat cerebral cortex (7). Recent studies have shown lithium inhibition of adenylate cyclase stimulated by forskolin in both membranal preparations and slices of rat cerebral cortex after both in vivo and ex vivo administration (7,8,9). Forskolin was originally proposed to stimulate adenylate cyclase by direct activation on the catalytic unit (10). Recently, it has been suggested that its site of action is the Gs protein (11,12). It is now recognized that two binding sites for forskolin exist: a low-affinity site on the catalytic unit and a high-affinity site located on Gs (13). All of these findings suggest a postreceptor site for lithium action. However, no differentiation has been made between Gs and the catalytic unit.

An important characteristic of G-proteins is their increased guanine nucleotide binding following agonist stimulation which in turn leads to their activation. Increased binding of radiolabelled guanine nucleotides induced by beta-adrenergic agonists has been shown in both membranal and purified reconstituted systems (16). We investigated the direct effect of lithium on Gs function by examining its effect on isoproterenol-induced increases in GTP-binding (17). Isoproterenol affects GTP maximal binding capacity with no effect on its binding affinity. This isoproterenol-induced increase in GTP binding can be abolished by propranolol. Lithium at therapeutically efficacious concentrations (0.4–1.2 mM) completely blocked isoproterenol effects on GTP binding to Gs with no effect on basal GTP binding.

The in vitro effects of lithium on Gs were substantiated by ex vivo experiments in chronically lithium-treated rats. Chronic treatment with daily i.p. injections of Li_2CO_3 for 21 days, reaching therapeutic blood levels (0.6–1.0 mM) prevented isoproterenol-induced increases in GTP binding capacity. Withdrawal of lithium for 48 hours resulted in full recovery of isoproterenol-induced increases in GTP binding.

Our results suggest Gs as the molecular site for lithium action. Lithium effects on this protein can explain the findings on the inhibitory effect of this drug on isoproterenol-, Gpp (NH)p-, and forskolin-stimulated adenylate cyclase activity. Lithium, found in this study (17) to inhibit G-protein function, is the first therapeutically used drug to be shown to interact with G-protein. Future studies on G-protein function and their interactions with psychoactive drugs may help to unravel the molecular mechanisms underlying affective disorders and their treatment.

However, biochemical effects of psychoactive drugs should be correlated with behavioral effects in animals. Otherwise, the biochemical effect could conceivably be an epiphenomenon unrelated to the drugs' clinical mode of action. We therefore attempted to find specific effects of lithium on rat behavior mediated via noradrenergic adenylate cyclase.

Salbutamol and clenbuterol have been reported to reduce rat activity in the first 3 to 5 minutes of exploratory behavior in an open field (18,19). This clenbuterol- or salbutamol-induced effect is preventable by prior treatment with propranolol (20). Selective central noradrenergic supersensitivity without peripheral noradrenergic supersensitivity was associated with enhanced responsiveness to clenbuterol (21). Thus, clenbuterol-induced hypoactivity may be mediated by central beta-2 adrenergic receptors. Chronic treatment of rats with imipramine or similar antidepressants led to a reduction in ligand-binding beta-adrenergic receptor number in the brain and a parallel subsensitivity to the behavioral effects of clenbuterol (18,22).

Lithium (Li) has been reported to inhibit noradrenergic adenylate cyclase distal to the beta-receptor (1) and thus functionally to accomplish a similar reduction in noradrenergic receptor function to imipramine-like antidepressants which reduce beta-receptor number on chronic treatment (3). We therefore decided to study the effect of chronic Li on clenbuterol-induced hypoactivity, in an attempt to find a specific behavioral effect of Li attributable to lithium's specific biochemical effect on beta-adrenergic adenylate cyclase.

Sabra rats (130–150 g) were maintained 5 to a cage at 20°C with 12 hour light/dark cycles. Rat food containing 0.15% LiCl was prepared by grinding regular rat pellets to a fine powder and mixing thoroughly with LiCl. Forty rats were

TABLE 7.1. Open field crossings in rats treated with chronic lithium followed by clenbuterol.

	Control	Clenbuterol only	Li-food only	Li-food + clenbuterol
Experiment 1*	43 ± 16	18 ± 11	32 ± 22	24 ± 14
($\overline{X}$ ± SD)	n = 20	n = 20	n = 20	n = 19
Experiment 2**	76 ± 19	39 ± 14	110 ± 23	30 ± 16
($\overline{X}$ ± SD)	n = 9	n = 10	n = 9	n = 10

*crossings/3 minutes
**crossings/5 minutes

divided into 4 groups as follows: 1) food with LiCl for approximately 3 weeks, followed by clenbuterol 0.2 mg/kg i.p.; 2) regular food for 3 weeks, followed by clenbuterol i.p.; 3) food with LiCl for 3 weeks, followed by saline 0.2 mg/kg i.p.; 4) regular food for 3 weeks, followed by saline i.p. Behavioral testing was done for an initial 3 minutes in an open field, one square meter, where rats were counted for a) crossing peripheral squares, b) crossing central squares, and c) rearing on hind legs. After behavioral testing, carotid blood was obtained from lithium-fed rats for determination of Li blood levels by flame photometry.

In a second experiment, larger rats (200 g) were fed 0.2% LiCl in food and treated as previously, except that observation lasted for 5 minutes.

Mean serum Li levels were 0.33 mM in experiment 1 and 0.55 mM in experiment 2. Since Li alone is known to depress spontaneous activity (23), a minimal dosage schedule of Li was chosen and blood levels were indeed low compared to those in previous experiments (1).

Table 7.1 illustrates the results. Clenbuterol induces a significant and consistent reduction in square crossings in the open field. Li only had a smaller but less consistent effect to reduce activity in experiment 1. Clenbuterol's effect to reduce activity was not significantly prevented in lithium-treated rats, despite a trend in this direction in experiment 1. (24.5 vs 18.3, t = 1.5ns). In experiment 2 there was an unusual effect of Li alone to increase crossings.

In these experiments chronic Li could not clearly prevent clenbuterol-induced hypoactivity and this contrasts with results previously reported for imipramine and similar antidepressants (18). In vivo Li inhibition of noradrenergic adenylate cyclase occurs in the rat brain at high Li levels, around 1.7 mM (1), compared with the human brain which appears more sensitive to Li inhibition (24). If clenbuterol-induced hypoactivity is indeed a centrally mediated behavior dependent on beta-2 adrenergic receptors, then Li inhibition of such receptors may not to occur in the rat at Li blood levels achieved in this experiment. Electroconvulsive shock, which has some similarities in clinical profile in affective disorders to Li (25) and which reduces B-adrenergic receptor number as measured by ligand binding in rat brain (26), did not cause a reduction in clenbuterol-induced hypoactivity (27). Thus, ECS and Li may cause a reduction in central B-adrenergic receptor function which is not measurable using clenbuterol-induced hypoactivity. Alternatively, the Li level may be the critical issue, and Li levels which are high enough to inhibit noradrenergic adenylate cyclase in the rat brain may depress spontaneous activity too much for clenbuterol-induced hypoactivity to be a useful test. These issues are similar to those in studies of effects of chronic Li feeding in rats on the hyperactive response to amphetamine (28), where dosage and blood levels are crucial variables for a drug such as Li with a very narrow therapeutic index.

A less specific but highly important "behavior" that an be used to test theories of lithium's mechanism of action is toxicity. Lithium (Li) is a drug well known for its narrow therapeutic index and toxicity or overdosage is a troublesome clinical problem (29). Severe overdosage has significant mortality (30). There is no known physiological antidote, such as physostigmine for anticholinergic toxicity or nalorphine for opiate overdosage. If the theory is correct that Li ameliorates mania via inhibition of noradrenergic adenylate cyclase in the brain, and Li toxicity represents generalized Li inhibition of numerous adenylate cyclases at higher than therapeutic Li concentrations (3) then forskolin, a diterpine agent capable of directly and powerfully stimulating adenylate cyclase via the catalytic site in numerous tissues (10) might be capable of reversing Li toxicity.

Another leading theory of Li action claims that Li ameliorates mania by inhibiting inositol-1-phosphatase (31), the enzyme that regenerates inositol for resynthesis to phosphatidylinositol (PI) in the PI cycle, and thus reducing function of PI-linked receptors. Since the PI cycle is of widespread importance, this theory views Li toxicity as due to a generalized inhibition of PI-linked receptors. If this theory is correct, then inositol, a simple sugar, might be capable of reversing Li toxicity by restoring the lithium-induced depletion of inositol. Inositol has been reported to reverse lithium-induced alternations in PI metabolism using in vitro systems (32).

We decided to test these two hypotheses in a mouse model of Li toxicity (33). Male mice, 25 to 40 gm, of Balb C strain, were used in all experiments. Li toxicity was induced by i.p. injection of 35 mM/kg LiCl in isotonic solution (approximately 7 cc volume). Mice were observed at frequent (½ h) intervals after the injection and the usual syndrome of Li toxicity was noted, including sedation, piloerection, diarrhea, tremor, myoclonic jerking, ataxia, and death.

Forskolin 250 mg was dissolved in 20 cc of ethanol + 5 cc water and given i.p. in a dosage of 6 mg/kg or 12 mg/kg. Meso-inositol (BDH Company) was dissolved 1 gm/5 cc water and given in dosages of 0.5 gm, 1.0 gm, 1.5 gm, or 2.0 gm per mouse, i.p. Table 7.2 illustrates the results.

Neither of the two proposed antidotes, both based on pathophysiological theories of lithium's mechanism, were effective in preventing Li toxicity. There is evidence that both compounds are able to penetrate the brain. Forskolin i.p. is reported to have behavioral effects (34) and in our experiments was also noted to cause sedation of those animals that received it. Inositol, even in lower doses than given here, has been found in our laboratory (Z. Kaplan, unpublished data) to exacerbate pilocarpine + Li induced seizures in rats, a finding perhaps due to

TABLE 7.2. Effect of forskolin or inositol on lithium toxicity.

Experiment	Toxic stimulus	Treatment	Dead after 24 hours
1	LiCl 35 mM/kg	Forskolin 6 mg/kg	20/40
		Saline	23/40
2	LiCl 35 mM/kg	Forskolin 12 mg/kg	10/10
		Saline	6/10
3	LiCl 35 mM/kg	Forskolin 6 mg/kg/hour	8/10
		Saline	3/5
4	LiCl 35 mM/kg	Saline	6/10
		Inositol	
		0.5 g/mouse	5/10
		1.0 g/mouse	7/10
		1.5 g/mouse	4/10
		2.0 g/mouse	6/10
5	LiCl 35 mM/kg	Saline	10/15
		Inositol 0.5 g every two hours	11/15

enhanced availability of inositol allowing increased generation by pilocarpine of diacyl glycerol and/or inositol triphosphate or other second messengers putatively involved in the seizures due to Li + pilocarpine (35).

Our results are disappointing clinically but could mean that both the adenylate cyclase theory of Li action and the phosphatidylinositol theory may be oversimplified or untrue. Newman and Belmaker (7) have found that Li can inhibit forskolin-induced rises in cyclic AMP in rat cortex, and thus even this direct adenylate activator may not act distal enough to the Li block. Alternatively, lithium's toxic effects may be mediated by pathophysiologic mechanisms different from its therapeutic effects.

Seizures induced by Li and pilocarpine have been attributed to Li effects on cholinergically mediated phosphatidylinositol hydrolysis. Honchar et al. (35) recently observed that treatment of rats with doses of lithium roughly equivalent to those used therapeutically in patients with affective disorders, prior to administration subconvulsive doses of cholinergic agonists, caused seizures. These findings were confirmed by others (36). Since phosphatidylinositol (PI) turnover is stimulated by cholinergic agents actings on muscarinic receptors, and since lithium has been shown to affect PI metabolism, it was hypothesized by Honchar et al. (35) that seizures induced by cholinergic agents in lithium-treated rats stem from altered phosphoinositide metabolism.

In recent years, evidence has accumulated supporting the existence of multiple subtypes of muscarinic-cholinergic receptors (37). It has been proposed that those sites which show a high degree of sensitivity to pirenzepine and which are labelled with high affinity by }³H{-pirenzepine are named M-1 receptors, and M-2 recep-

TABLE 7.3. Blockage of Li-pilocarpine seizures by muscarinic antagonists with varying M1 receptor selectivity.

Antagonist	Affinity to M1 subtype (nM)	Affinity to M2 subtype (nM)	I 50 (mg/kg)
AD FX-116*	670	100	2.5
Biperiden	2.7	38	0.5
Procyclidin	12	94	0.75
Methixen	17	17	0.75

*Data for ADFX 116 were taken from ref. 37, and for the other anticholinergics from ref. 38.

tors are those with low affinity for pirenzepine. M-2 receptors have been proposed to be coupled to inhibition of adenylate cyclase and M-1 receptors to be coupled to PI hydrolysis. G-proteins are important in both of these biochemical systems.

In order to attempt to confirm the Li-pilocarpine seizures as a possible behavioral model of Li effects, selective muscarinic antagonists for both M-1 and M-2 subtypes as well as nonselective antagonists were compared in their efficacy to prevent seizures and subsequent lethality induced by cholinergic agonists in lithium-treated rats.

We used the antiparkinsonian drugs biperiden and procyclidine as prototypes of M1-selective antagonists and methixen as a nonselective muscarinic antagonist (38). AD FX116 is used as a selective M2 antagonist (39). Several concentrations of each drug were tested for their ability to block the seizures induced by the simultaneous injection of lithium and pilocarpine. Dosages that blocked seizures in 50% of the animals were determined (Table 7.3).

Biperiden, procyclidine, and methixen are respectively five, three, and three times more efficient than ADFX-116 in blocking seizures (Table 7.3). However, their affinity towards the M1-subtype of muscarinic receptors is 40 to 250 times higher than that of AD FX116. Thus, it is unlikely that Li-pilocarpine induced seizures are mediated by M1 receptors. Procycliden has about the same affinity as AD FX116 for M2 receptors but is three times more potent in blocking Li-pilocarpine seizures. Thus a specific M2 mediated mechanism of Li-pilocarpine seizures is also unlikely. If Li-pilocarpine seizures cannot be clearly ascribed to a PI-linked receptor or to an adenylate cyclase linked receptor, then this behavioral model of Li effects is less useful.

In summary, clear effects of Li to inhibit agonist-induced increases in GTP binding to G-protein in rat brain have not yet found clear behavioral correlates. Such behavioral correlates must be sought if the link between lithium's effects on behavior and its effect on biochemistry is to be clarified.

References

1. Ebstein, R.P., Hermoni, M., & Belmaker, R.H. (1980). The effect of lithium on noradrenaline-induced cyclic AMP accumulation in rat brain; inhibition after chronic treatment and absence of supersensitivity. *J. Pharmacol. Exp. Ther., 213,* 161–167.
2. Newman, M.E., Lichtenberg, P., & Belmaker, R.H. (1985). Effects of lithium *in vitro*

on noradrenaline-induced cyclic AMP accumulation in rat cortical slices after reserpine-induced supersensitivity. *Neuropharmacol. 24*, 353–355.

3. Belmaker, R.H. (1981). Receptors, adenylate cyclase, depression and lithium. *Biol. Psychiat., 16*, 333–350.

4. Rodbell, M. (1980). The role of hormone receptors and GTP- regulatory proteins in membrane transduction. *Nature, 284*, 17–19.

5. Gilman, A.G. (1984). Guanine nucleotide regulatory proteins and dual control of adenylate cyclase. *J. Clin. Invest., 73*, 1–4.

6. Maggi, A., & Enna, S.J. (1980). Regional alternations in rat brain neurotransmitter systems following chronic lithium treatment. *J. Neurochem., 34*, 888–892.

7. Newman, M.E. & Belmaker, R.H. (1987). Effects of lithium in vitro and ex vivo on components of the adenylate cyclase system in membranes from the cerebral cortex of the rat. *Neuropharmacol., 26*, 211–217.

8. Andersen, P.H., & Geisler, A. (1984). Lithium inhibition of forskolin-stimulated adenylate cyclase. *Neuropsychobiol., 12*, 1–3.

9. Andersen, P.H., Klysner, R., & Geisler, A. (1984a) Forskolin-stimulated adenylate cyclase activity in rat cerebral cortex following chronic treatment with psychotropic drugs. *Acta Pharmacol. Toxicol., 55*, 278–282.

10. Seamon, K.B., Padgett, W., & Daly, J.W. (1981). Forskolin: Unique diterpene activator of adenylate cyclase in membranes and in intact cells. *Proc. Natl. Acad. Sci.* (USA), *78*, 3363–3367.

11. Krall, J.F. (1984). A kinetic analysis of activation of smooth muscle adenylate cyclase by forskolin. *Arch. Biochem. Biophys., 229*, 492–497.

12. Barber, R., Goka, T.J. (1985). Adenylate cyclase activity as a function of forskolin concentration. *J. Cyclic Nucl. Protein Phosph. Res., 10*, 23–29.

13. Morris, S.A., & Bilezikian, J.P. (1983). Evidence that forskolin activates turkey erythrocyte adenylate cyclase through a noncatalytic site. *Arch. Biochem. Biophys., 220*, 628–636.

14. Barovsky, K., Pedone, C., & Brooker, G. (1984). Distinct mechanisms of forskolin-stimulated cyclic AMP accumulation and forskolin-potentiated hormone responses in C6-2B cells. *Molec. Pharmacol., 25*, 256–260.

15. Seamon, K.B., & Daly, J.W. (1986). Forskolin. Its biological and chemical properties. *Adv. Cyclic Nucl. Protein Phosh. Res., 20*, 1–149.

16. Cassel, D., & Selinger, Z. (1978). Mechanism of adenylate cyclase activation through the beta-adrenergic receptor: Catecholamine-induced displacement of bound GDP by GTP. *Proc. Natl. Acad. Sci.* (USA), *75*, 4155–4159.

17. Avissar, S., Schreiber, G., Danon, A., et al. (1988). Lithium inhibits adrenergic and cholinergic increases in GTP binding in rat cortex. *Nature, 331*, 440–442.

18. Mogilnicka, E. (1982). The effects of acute and repeated treatment with salbutamol, a B-adrenoceptor agonist, on clonidine-induced hypoactivity in rats. *J. Neural Transm., 53*, 117–126.

19. Ortmann, R., Meisburger, J.G. & Mogilnicka, E. (1985). Effect of B-adrenoceptor agonists on apomorphine-induced turning in rats. *J. Neural Transm., 61*, 43–53.

20. Nowack, G. & Mogilnicka, E. (1985). The influence of B-agonists on normal and clonidine-decelerated noradrenaline turnover in the rat brain. *Pol. J. Pharmacol. Pharm., 37*, 765–772.

21. Dooley, D.J., Mogilnicka, E., Delini-Stula, A., et al. (1983). Functional supersensitivity to adrenergic agonists in the rat after DSP-4, a selective noradrenergic neurotoxin. *Psychopharmacol., 81*, 1–5.

22. Mogilnicka, E., Klimek, V., Nowak, G., et al. (1985). Adaptive and differential

changes on B- and A_2-adrenoreceptors mediated hyperthermia after chronic treatment with antidepressant drugs in the rat kept at high ambient temperature. *J. Neural Transm.*, *63*, 237–246.

23. Lerer, B., Globus, M., Brik, E., et al. (1984). Effect of treatment and withdrawal from chronic lithium in rats on stimulant-induced responses. *Neuropsychobiol.*, *11*, 28–32.

24. Newman, M., Klein, E., Birmaher, B., et al. (1983). Lithium at therapeutic concentrations inhibits human brain noradrenaline-sensitive cyclic AMP accumulation. *Brain Res.*, *278*, 380–381.

25. Lerer, B., Stanely, M. & Belmaker, R.H. (1984). ECT and lithium: Parallels and contrasts in receptor mechanisms. In B. Lerer, R.D. Weiner, & R.H. Belmaker (Eds.), ECT: Basic mechanism (pp. 67–78). London: John Libbey.

26. Belmaker, R.H., Lerer, B., Bannet, J., et al. (1981). The effect of electroconvulsive shock at a clinically equivalent schedule on rat cortical B-adrenoreceptors. *J. Pharm. Pharmacol.*, *34*, 275.

27. Kleinrok, Z. & Wielosz, M. (1985). A role of some central neurotransmitter systems in the mechanism of electroconvulsive shock action. *Pol. J. Pharmacol. Pharm.*, *37*, 745–752.

28. Berggren, U. (1985). Effects of chronic lithium treatment on brain monoamine metabolism and amphetamine-induced locomotor stimulation in rats. *J. Neural Transm.*, *64*, 239–250.

29. Greenblatt, D.J., & Shader, R.I. (1985). Psychotropic drug overdosage. In R.I. Shader, (Ed.), *Manual of psychiatric therapeutics*. Boston: Little, Brown and Co..

30. Applebaum, P.S., Shader, R.I., Funkenstein, H.H., et al. (1979). Difficulties in the clinical diagnosis of lithium toxicity. *Am. J. Psychiat.*, *136*, 1212–1213.

31. Berridge, M.J., Downes, C.P., & Hawley, M.R. (1982). Lithium amplifies agonist-dependent phosphatidylinositol responses in brain and salivary glands. *Biochem. J.*, *206*, 587–595.

32. Downes, C.P., & Stone, M.A. (1986). Lithium-induced reduction in intracellular inositol supply in cholinergically stimulated parotid gland. *Biochem. J.*, *234*, 199–204.

33. Zohar, J., Spiro, D., Novack, A., et al. (1982). Lack of benefit from magnesium in lithium toxicity. *Nueropsychobiol.*, *8*, 10–11.

34. Wachtel, H., & Loschmann P.-A. (1986). Effects of forskolin and cyclic nucleotides in animal models predictive of antidepressant activity: Interactions with rolipram. *Psychopharmacol.*, *90*, 430–435.

35. Honchar, M.P., Olney, J.W., & Sherman, W.R. (1983). Systemic cholinergic agents induce seizures and brain damage in lithium-treated rats. *Science, 220*, 323–325.

36. Jope, R.S., Morrisett, R.A., & Snead, O.C. (1986). Characterization of lithium potentation of pilocarpine induced status epilepticus in rats. *Exp. Neurol, 91*, 471–480.

37. Hammer, R., Berrie, C.P., Birdsall, N.J.M., et al. (1980). Pirenzepine distinguishes between different subclasses of muscarinic receptors. *Nature* (Lond.), *283*, 90–94.

38. Schreiber, G., Avissar, S., & Belmaker, T.H. Differences in the selectivity of antiparkinsonian-anticholinergic drugs towards muscarinic receptor subclasses: M1-selective versus non-selective binding patterns, submitted for publication.

39. Hammer, R., Giraldo, E., Schiavi, G.B., et al. (1986). Binding profile of a novel cardioselective muscarinic receptor antagonist, AFDX-116, to membranes of peripheral tissues and brain in the rat. *Life Sci.*, *38*, 1653–1162.

8
The Effect of Lithium on Inositol Phosphate Metabolism

C. Ian Ragan

Introduction

As a tool for research on the molecular mechanism of manic depression, it is hard to envisage anything less promising than lithium. Compared with most drugs, it is of very low potency and there are a large number of biochemical processes which could be affected by lithium at therapeutically relevant concentrations (1,2). It is therefore extremely difficult to decide which might be pertinent to its action. Lithium has also remained the most effective treatment for nearly 40 years, attesting to its unique properties and the obvious impossibility of designing more potent, specific, and efficacious analogues through the traditional principles and methodology of drug design. How then are we to approach the problem of the mechanism by which lithium acts? Obviously we must look for likely targets which are modulated by lithium within its therapeutic dose range and from which we can construct a plausible case that this modulation underlies both the antimanic and antidepressant actions of lithium. This is extremely difficult and there is by no means a consensus on the most likely target. This chapter reviews the action of lithium on a target which fulfills the preceding criteria reasonably well, that is, the phosphoinositide cycle. The discussion covers two main issues; first, whether the actions of lithium on the cycle can be explained by inhibition of a single step in the pathway, for example, the inositol monophosphatase enzyme; and second, whether interference with the phosphoinositide cycle provides a convincing explanation for the therapeutic action of lithium.

The Phosphoinositide Cycle

The scheme of Figure 8.1 provides an incomplete but noncontroversial outline of the phosphoinositide cycle. The key step is the phosphodiesteratic cleavage of phosphatidylinositol 4,5-bisphosphate (PIP_2) by phospholipase C, in response to

---

*The abbreviated formulae for inositol phosphates used in this article refer to the D-enantiomer of esters of *myo*-inositol. When the particular isomer is not specified (e.g., Ins(1)P

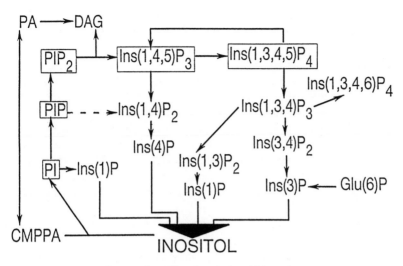

FIGURE 8.1. The phosphoinositide cycle.

an appropriate agonist, giving rise to Ins(1,4,5)P$_3$* and diacylglycerol (DAG) (3,4). The role of these two substances in regulating intracellular Ca^{2+} and protein kinase C activity, respectively, is well established (5), although the exact part played by Ins(1,4,5)P$_3$ and Ins(1,3,4,5)P$_4$ in the influx of extracellular Ca^{2+} and the reversible release of Ca^{2+} from intracellular stores remains unclear and certainly outside the scope of this chapter. While the function of the various inositol poly-phosphates ranges from contentious to unknown, the pathways by which they are formed are now becoming clearer after a flurry of activity over the past few years. Thus, Ins(1,4,5)P$_3$ is dephosphorylated to inositol via Ins (1,4)P$_2$ (6–9) and Ins(4)P (10–14) and earlier reports that Ins(1)P was also an intermediate (15,16) are now acknowledged as incorrect (17). Ins(1,4,5)P$_3$ can also be phosphorylated by a 3-kinase to Ins(1,3,4,5)P$_4$ (18,19). Until recently, removal of Ins(1,3,4,5)P$_4$ was thought to be exclusively via a 5-phosphatase giving rise to Ins(1,3,4)P$_3$ (8,20–22). However, the discovery of 3-phosphatases in erythrocyte membranes (23), brain cytosol (24) and rat basophilic leukemic cell membranes (25) which return Ins(1,3,4,5)P$_4$ to Ins(1,4,5)P$_3$ has complicated the issue. The existence of a futile cycle interconnecting Ins(1,4,5)P$_3$ and Ins (1,3,4,5)P$_4$ as a regulatory device would not be surprising if Ins(1,3,4,5)P$_4$ or one of its metabolites were shown convinc-ingly to have some important function in the cell. The metabolism of Ins(1,3,4)P$_3$ is complex since it can be dephosphorylated via either Ins(1,3)P$_2$ or Ins(3,4)P$_2$ to inositol or can be phosphorylated to Ins(1,3,4,6)P$_4$, (21,30–31) a precursor for

---

and Ins(4)P) the compound is denoted thus: InsP$_1$, InsP$_2$, etc. The plane of symmetry in the *myo*-inositol molecule means that alternative formulae are in current use. Thus D-Ins(3)P is identical with L-Ins(1)P and D-Ins(3,4,5,6,)P$_4$ is identical with L-Ins(1,4,5,6,)P$_4$. How-ever, the exclusive use of D-enantiomer numbering makes the metabolic derivation of the compound much clearer.

Ins(1,3,4,5,6)P$_5$ (32). The richness of the metabolism in this region of the cycle has given rise to speculation that one or more of these metabolites is important and perhaps the claim that Ins(1,3,4,5,6)P$_5$ acts extracellularly as a neurotransmitter or modulator provides support for this idea (33).

The scheme shown is an oversimplification in many ways. First, the metabolites shown are those that respond rapidly to cell stimulation. Others, such as Ins(3,4,5,6)P$_4$ (34) and Ins(1,3,4,5,6)P$_5$ (32) respond poorly or slowly to cell stimulation and their role in signal transduction within the cell is doubtful. Indeed, Ins(3,4,5,6)P$_4$ is not synthesized from any known "agonist-sensitive" phosphoinositide cycle intermediate (34). Second, the details of the pathway may differ from one cell type to another. For example, the extent to which direct phosphatidylinositol 4-phosphate (PIP) and phosphatidylinositol (PI) hydrolysis contribute to the inositol phosphate pools is largely unknown, but is likely to vary from cell type to cell type. Thus, the amount of Ins(4)P produced in stimulated cells is clearly a marker for cleavage of PIP and PIP$_2$ (Figure 8.1), and in some cells (e.g., cultured pituitary cells) (35), Ins(4)P responds to cell stimulation to a greater extent than Ins(1)P. However, in other cells such as in brain (36) and platelets (37), Ins(1)P is the major monophosphate accumulating in the presence of agonist, and may arise from direct PI hydrolysis. However, the distinction between Ins(1)P and its enantiomer Ins(3)P is rarely made and in cells with an active Ins(1,3,4,5)P$_4$ branch of the cycle, both Ins(1)P and Ins(3)P could arise primarily through dephosphorylation of Ins(1,3,4,5)P$_4$ (38). In cerebral cortical slices, a substantial fraction of Ins(1,4,5)P$_3$ metabolism is believed to take place via Ins(1,3,4,5)P$_4$ (39), which may account for the larger accumulation of Ins(1)P rather than Ins(4)P. Similarly, the relative rate of Ins(1,3,4)P$_3$ degradation via Ins(1,3)P$_2$ and Ins(3,4)P$_2$ is tissue dependent. For example, Ins(3,4)P$_2$ is the only product formed in stimulated polymorphonuclear leukocytes (40) and is the major product of Ins(1,3,4)P$_3$ hydrolysis by all bovine tissue homogenates tested except for brain, where Ins(1,3)P$_2$ predominates (14).

The significance of these variations is largely unknown. Obviously direct phospholipase C action on PI and PIP offers a means of producing DAG without directly affecting Ca^{2+} mobilization via Ins(1,4,5)P$_3$, but the consequences of the other metabolic variations just noted remain obscure while roles for Ins(1,3,4,5)P$_4$ and its dephosphorylation products are undecided.

The scheme for Figure 8.1 omits any mention of inositol cyclic phosphates. The evidence is quite convincing that cIns(1:2)P (41), cIns(1:2,4)P$_2$ (42), and cIns(1:2,4,5)P$_3$ (42–45) are formed in vivo by phospholipase C action on PI, PIP, and PIP$_2$ respectively. Indeed, the formation of the former in carbachol-stimulated pancreatic minilobules (41) provides good evidence that direct phosphodiesteratic cleavage of PI can occur in some cells. The role of these cyclic phosphates, in particular cIns(1:2,4,5)P$_3$, is still unsure. The rate of formation and degradation of the cyclic trisphosphate is much slower than Ins(1,4,5)P$_3$, for example, in pancreatic cells (44), platelets (43), and parotid acinar cells (45). The cyclic and noncyclic trisphosphates are approximately equipotent in stimulating Ca^{2+} release (46), but the rate of synthesis of cIns(1:2,4,5)P$_3$ is too slow for it to be the primary regulator of cytosolic Ca^{2+} (44–45), even though it may have some role in prolonging the Ca^{2+} signal following long-term stimulation.

The Effect of Lithium on the Phosphoinositide Cycle In Vivo

The first observations of lithium action on the phosphoinositide cycle were made by Allison and Stewart in 1971 (47), who showed that following a single s.c. dose of lithium (10 mmol/kg), inositol levels in rat cerebral cortex decreased by as much as 35%. Early experiments also demonstrated that the decline in inositol in cortex was dependent on muscarinic cholinergic stimulation and that similar decreases could be found in several other brain regions (hypothalamus, hippocampus, caudate) but not in cerebellum or corpus callosum (48–49).

Later work of Allison, Sherman, and co-workers showed that the decline in inositol was matched by a parallel increase in Ins(1)P (Figure 8.2) which could be elevated 40-fold by lithium (49–51). This experiment clearly points to inhibition by lithium of the enzyme that dephosphorylates Ins(1)P to inositol and this was demonstrated in vitro by Hallcher and Sherman (52) and others as detailed in a later section.

The key observations relating to the cortical effect of lithium were as follows. First, the effect is mainly exerted on Ins(1)P (30). Ins(3)P and Ins(4)P also rise sharply, but the concentrations attained are an order of magnitude lower than that of Ins(1)P. Second, the extent of inositol depletion and Ins(1)P elevation is greatly potentiated by the muscarinic agonist, pilocarpine and, moreover, the duration of limbic seizures caused by the combination of lithium and pilocarpine is strongly correlated with the Ins(1)P level (49). Only under extreme conditions, with high doses of both lithium and pilocarpine, can depletion of phosphoinositides be detected. Thus, conversion of up to 60% of the free inositol into Ins(1)P had surprisingly little effect on the levels of PI, PIP, and PIP_2 (49). The relevance of this to the theory that lithium exerts its antimanic effect through the PI cycle is considered later in this chapter.

The effects of lithium on Ins(1)P levels are maintained in rats given dietary lithium over a three-week period (49,51). Cortical lithium levels fluctuate but the concentration correlates very well with the level of Ins(1)P providing no evidence that chronic lithium results in compensatory changes in enzyme levels to restore the normal metabolite levels. This result contrasts with a later study showing induction of inositol monophosphatase following chronic lithium treatment (53) and independent corroboration would be desirable. Of particular interest is the apparent tissue specificity of the lithium effect. Thus, while cortical lithium levels reach only one third of those in kidney, kidney inositol monophosphate levels reach only 20% of the cortical concentration (51). Moreover, in kidney, 70% of this is Ins(3)P, formed either by de novo synthesis from Glu(6)P or from hydrolysis of $Ins(3,4)P_2$. Whatever the route of production, the small increase in inositol monophosphates in kidney and other peripheral tissues (testis, liver) provides a nice biochemical correlate of lithium's central action and

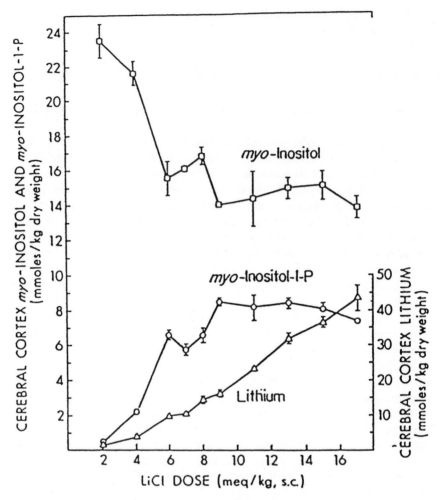

FIGURE 8.2. Elevation of Ins(1)P by systemic lithium in rat cerebral cortex. Subcutaneous injection of LiCl was done 20 hours prior to assay of cortical Ins(1)P and Ins levels. [Reproduced from ref. 51, with permission of Raven Press.]

suggests that PI cycle flux in these peripheral tissues is small compared with the capacity of inositol monophosphatase (see following).

Much has been made of the inability of inositol to cross the blood brain barrier and the consequent inability of the brain to restore the lithium-depleted inositol pool from the diet (54,55). It is difficult to assess the significance of this to the apparent central selectivity of lithium since peripheral tissues do not show much change in Ins(1)P either and we do not know which is the most important. Some effects of lithium, both biochemical and pharmacological, are reversed by inositol (see later), but this has not been systematically explored. Furthermore, since depletion of phosphoinositides can only be demonstrated with brain-damaging doses of lithium and pilocarpine (49), the idea that selective depletion of PIP_2 through reduction in free inositol is a major factor in the therapeutic action of lithium has yet to receive convincing experimental support.

The experiments performed by Sherman and colleagues are important because they demonstrate that alterations in PI metabolism can occur in vivo at therapeutically relevant concentrations of lithium. However, the technical difficulty of measuring the concentrations of inositol phosphates, particularly the polyphosphates, clearly restricts the range of experimentation and less heroic individuals prefer the comparative simplicity of tissue slices or cultured cells.

An interesting halfway house between in vivo and in vitro has been examined by several groups, that is, PI turnover in cortical slices from lithium-treated animals. While Sherman and colleagues had been unable to demonstrate any changes in muscarinic receptor response in vivo after chronic lithium, effects in cortical slices are readily observed.

Kendall & Nahorski showed that acute or chronic lithium treatment causes a significant reduction in the ability of carbachol, histamine, 5-HT, or K^+-evoked depolarization to increase inositol monophosphate levels in rat cortical slices. Noradrenergic responses are more resistant to lithium treatment and a decreased response is only observed following chronic treatment, in agreement with the work of Casebolt and Jope (57). However, in another study (58), acute lithium was found to potentiate the noradrenergic response. Since the former study employed a high saturating concentration of lithium in the cortical slice assay, while the second used much lower, it is possible that the potentiation resulted from residual lithium in the slice preparation.

The inability of inositol to reverse the lithium-induced decrease in PI turnover, the lack of any effect of acute lithium on phospholipase C activity and inhibition of PI turnover linked to K^+-evoked depolarization point to an indirect effect of lithium (56). A possibility suggested by Kendall and Nahorski is that lithium increases DAG levels (as has been demonstrated in GH_3 cells) (59) and that resulting protein kinase C activation is responsible for the suppression of PI turnover, perhaps via a G protein or other regulatory device. In keeping with this, Labarca et al. (60) reported inhibition of carbachol-stimulated PI turnover in hippocampal slices by phorbol esters.

The mechanism by which lithium might elevate DAG is considered in the next section on in vitro experiments.

The Effects of Lithium on the Phosphoinositide Cycle In Vitro

The amplification of agonist-induced accumulation of inositol monophosphates in tissue slices and cells by lithium was first reported by Berridge and co-workers in 1982 (54) and was the crucial observation in their proposal that this might be the explanation for the therapeutic action of lithium. Since then, lithium has become a standard tool in many experiments on PI turnover in a wide variety of cell types. The following section is therefore restricted to experiments on cortical slices and experiments which address the nature of the lithium effect directly.

The ability of neurotransmitters to stimulate PI turnover in cortical slices has been reported in several publications. The most thorough studies of the effect of lithium are those of Batty and Nahorski (61,62) in rat and Whitworth and Kendall (39) in mouse. In rat, lithium causes increases in inositol monophosphates and inositol bisphosphates with ED_{50} values of 0.5 mM and 4 mM respectively (61). The former effect is undoubtedly a result of monophosphatase inhibition while the effect on bisphosphate concentration is most likely due to inhibition of $Ins(1,4)P_2$ 1-phosphatase as described in the next section. In mouse, large increases in $InsP_1$ and $InsP_2$ caused by lithium are also found with noradrenergic, histaminergic, serotonergic, or KCl stimulation (39). However, in the absence of an agonist, the effect of lithium is much weaker; at the respective ED_{50} concentration of lithium, Batty and Nahorski (61) found no significant elevation of $InsP_1$ or $InsP_2$ in the absence of an agonist.

Effects of $Ins(1,4,5)P_3$ and $Ins(1,3,4)P_3$ are less clear-cut. The earlier report of a marked decline in $InsP_3$ (61) was complicated by the presence in the "$InsP_3$" fraction of not only the two isomers but also $Ins(1,3,4,5)P_4$. As shown later by the same authors, carbachol-induced accumulation of $Ins(1,4,5)P_3$ is little affected by lithium in rat while $Ins(1,3,4)P_3$ accumulation is potentiated by lithium only over the first 15 minutes of stimulation (62). This could be explained by inhibition by lithium of $Ins(1,3,4)P_3$ breakdown (next section). On the other hand, in mouse cortical slices (39), total IP_3 levels are if anything decreased by lithium 10 minutes after carbachol stimulation, while marked increases are observed using histamine or noradrenaline as the agonist. $Ins(1,3,4,5)P_4$ is the intermediate in the pathway between the two trisphosphates, and the effects of lithium on this inositol polyphosphate are perhaps the most interesting and least understood. In rat, carbachol produces a rapid and sustained (>40 min) increase in $Ins(1,3,4,5)P_4$ (62). In the presence of lithium, however, the increase is sustained for only 10 minutes before falling rapidly over the next 10 minutes (Figure 8.3). Similarly, in the mouse (39), $Ins(1,3,4,5)P_4$ increases over a 5-minute period in the presence or absence of lithium, but thereafter declines when lithium is present. Lithium has no effect on the $Ins(1,3,4,5)P_4$ level in the absence of an agonist, but actually increases the $Ins(1,3,4,5)P_4$ concentration when histamine, noradrenaline, 5-HT, or KCl is used to stimulate the slice. However, carbachol is by far the most powerful agonist, producing $Ins(1,3,4,5)P_4$ levels 3- to 6-fold greater than those produced by other stimulants. Whitworth and Kendall (39) noted that $InsP_1$ accumulation is nevertheless very similar with different agonists and suggested that cells with muscarinic receptors might contain a more active $Ins(1,4,5)P_3$ kinase activity. In mouse cortical slices, the ED_{50} for inhibition of the $Ins(1,3,4,5)P_4$ response is less than 0.1 mM, a concentration insufficient, for example, to influence $Ins(1,4)P_2$ but enough to cause an appreciable elevation of $InsP_1$. The response of other cell types and tissues to agonists and lithium is similar in some regards to that encountered in cortical slices but by no means identical.

Potentiation of agonist-induced $InsP_1$ and $InsP_2$ accumulation by lithium has been reported in pituitary GH_3 cells (59), parotid acinar cells (63), platelets (64),

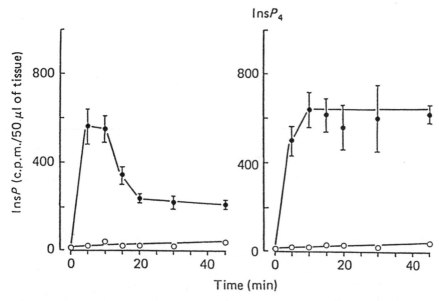

FIGURE 8.3. Effect of LiCl on the accumulation of [³H] Ins(1,3,4,5)P₄ stimulated by carbachol in rat cerebral cortical slices. Cortical slices were labelled with [³H] inositol for 30
minutes and stimulated with carbachol at zero time in the absence (right-hand graph) or
presence (left-hand graph) of 5 mM LiCl added 5 minutes previously. Open circles − no
carbachol; closed circles − plus 1 mM carbachol. [Reprinted by permission from Biochem
J, 247, pp. 797–800, copyright (c) 1987 The Biochemical Society, London.]

adrenal glomerulosa cells (10), and hepatocytes (65), for example. Increased
Ins(1,3,4)P₃ (or InsP₃) accumulation is also a common feature, but the attenuat-
ing effect of lithium on Ins(1,3,4,5)P₄ production has not been reported in other
tissues. For example, there is no effect of lithium on Ins(1,3,4,5)P₄ accumulation
in hepatocytes (66).

The mechanism of the effects of lithium on Ins(1,3,4,5)P₄ is obscure. It is
certainly not a result of any direct inhibition of the 3-kinase or activation of
the 5-phosphatase, and other suggestions are considered in the last section of
this chapter. One effect of lithium which may nevertheless be relevant to the
Ins(1,3,4,5)P₄ issue is the enhancement of cytidine monophosphorylphosphati-
date (CMPPA). This was most thoroughly studied in parotid acinar cells (63)
although very similar results can be obtained with cortical slices (67). In the
parotid study, there was a very marked elevation of CMPPA by lithium in the
presence but not the absence of carbachol, similar to the IP₁ response. Lithium
in the presence of carbachol produces a marked decrease in the amount of PI
present and both effects are fully reversed by inositol. These effects are obviously
a result of the decrease in inositol caused by blockade of monophosphatase and
CMPPA is a sensitive marker for this as cosubstrate with inositol for PI synthesis.
Despite the change in PI level, the concentration and rate of synthesis of PIP₂ are

TABLE 8.1. Properties of inositol monophosphatase.

Molecular	Native M_r	46,000–59,000 (52,68,71)
	Subunit M_r	29,000–31,000 (68–71)
Catalytic	K_m (Ins(1)P)	0.10–0.16 mM (52,68,71)
	K_m (Ins(4)P)	0.19 mM (69)
	K_m (β-glycerophosphate)	0.38–1.0 mM (68,70–71)
	K_m (2'-AMP)	0.19 (68), 0.58 mM (70)
	K_m (Mg^{2+})	1.0 (52), 2.0 mM (69)
	K_i (Li, versus Ins (1)P)	0.3–1.0 mM (68,70)
	K_i (Li, versus Ins(4)P)	0.26 mM (69)
	Vmax (Ins(1)P)	9.5–25 μmol/min/mg (68–70)

Values are for the crude (52) and purified bovine brain enzyme (69–71), or for the purified rat brain enzyme (68).

not altered and lithium does not reduce the secretory response of these cells. However, Downes and Stone (63) noted that lithium significantly enhances the increase in phosphatidic acid (PA) following carbachol treatment and since this was also reversed by inositol, they suggested that it might arise from CMPPA. Since PA can be dephosphorylated to DAG, an explanation was provided for the lithium-induced rise in DAG in GH_3 cells reported by Drummond and Raeburn (59). A mechanism is therefore provided through which lithium might cause an activation of protein kinase C and this topic is taken up again in the final section of this chapter.

Inhibition of Inositol Phosphatases by Lithium

Inositol Monophosphatase

The dramatic increases in inositol monophosphates and decrease in inositol caused by lithium both in vivo and in vitro clearly demonstrate that the inositol monophosphatase enzyme is a major target for lithium. The enzyme has been purified from rat (68) and bovine (69–71) brain and there is good agreement as to its general properties which are summarized in Table 8.1. The enzyme is a homodimer of two 30 kDa subunits, and catalyzes the Mg^{2+}-dependent hydrolysis of all *myo*-inositol monophosphates except Ins(2)P. In addition, a number of noninositol containing monophosphates are substrates, e.g., 2'-AMP, 2'-GMP, α- and β-glycerophosphates. However, inositol bis and polyphosphates are not hydrolyzed at all. Many divalent cations are quite potent inhibitors, particularly Zn^{2+} (K_i 2 μM) and these act as competitive inhibitors of Mg^{2+}. Lithium on the other hand is an uncompetitive inhibitor as first shown by Hallcher and Sherman (52), and amply confirmed by other workers using the purified enzyme (68–69). Confusion over whether Ins(4)P hydrolysis was lithium-sensitive has now been resolved (12) and it is agreed that Ins(4)P is hydrolyzed by the same enzyme as Ins(1)P and Ins(3)P as first demonstrated by Gee et al. (69). Ins(4)P hydrolysis

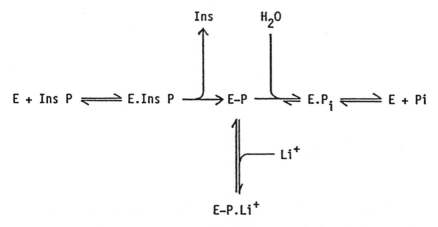

FIGURE 8.4. Mechanism of inositol monophosphatase. E.InsP and E.P$_i$ indicate Michaelis complexes while E-P denotes a covalent phosphoryl enzyme intermediate. Irreversibility of E-P formation is shown by the inability of inositol to inhibit the enzyme. The reversibility of the hydrolysis of E-P to E.P$_i$ is unknown.

is apparently more sensitive to lithium than that of Ins(1)P but this is not unexpected for an uncompetitive inhibitor. The simplest explanation for such inhibition is that the inhibitor binds only to the enzyme-substrate complex and therefore the K_i is likely to vary with the nature of the substrate. Indeed this so, with the K_i for lithium varying from as low as 0.26 mM with Ins(4)P to as high as 3.0 mM with β-glycerophosphate as substrate. However, Shute et al. (72) have proposed a different mechanism based on the ability of lithium-inhibited enzyme to catalyze a stoichiometric hydrolysis of Ins(1)P releasing inositol but not P$_i$. They propose therefore that lithium interacts with a phosphoryl-enzyme intermediate preventing its hydrolysis by water as shown in Figure 8.4. Despite the proposal that lithium binds to an intermediate common to hydrolysis of any substrate, and should therefore have a substrate-independent K_i, theoretical analysis shows that what is measured by standard steady-state kinetics is an apparent K_i which depends on the magnitude of the rate constant for formation of the phosphoryl-enzyme intermediate. Thus, the apparent K_i will still vary with the nature of the substrate. Uncompetitive inhibitors are not common pharmacological agents and the property that inhibition by such a compound becomes more marked as the substrate concentration rises is counterintuitive for those weaned on traditional competitive inhibitors. This has been partly responsible for conflicting literature reports about the lithium sensitivity of inositol monophosphate hydrolysis. As with competitive inhibitors, prediction of the metabolic consequences of an uncompetitive agent requires more than a knowledge of the K_i (or apparent K_i). The Km and concentration of the substrate are also needed. The relationship between these two quantities can be expressed as the degree of saturation of the enzyme which is the same as the fraction of the maximum flux at which it is operating. In a very illuminating paper, Cornish-Bowden (73) explored

TABLE 8.2. The effect of lithium on Ins(1)P levels.

Li$^+$	Ins(1)P (mM)	Fold increase in Ins(1)P by Li$^+$	v/V$_{max}$
−	0.040		0.268
+	0.075	1.9	
−	0.080		0.381
+	0.21	2.6	
−	0.10		0.435
+	0.43	4.3	
−	0.12		0.480
+	1.56	13	

Basal concentration of Ins(1)P in rat cerebral cortex is approx. 0.04 mM (50), and the lithium concentration was assumed to be the same as the K$_i$ for inhibition of monophosphatase (0.3–1 mM). v/V$_{max}$ was calculated from the assumed Ins(1)P concentration in the absence of Li$^+$ and K$_m$ for Ins(1)P of 0.13 mM. The Ins(1)P concentration in the presence of Li$^+$ was then calculated assuming that flux was unaltered by Li$^+$, i.e., the rise in Ins(1)P required to overcome blockade was determined. With increasing v/V$_{max}$ (or higher starting Ins(1)P concentration) greater increases in Ins(1)P are required to overcome the inhibition.

the theoretical effect of uncompetitive inhibitors on flux through metabolic pathways. He showed that competitive or uncompetitive inhibitors of an enzyme reaction will lead to an increase in substrate concentration which is much greater for steps in the pathway with a low flux control coefficient (i.e., a low dependence of overall velocity on the concentration of the particular enzyme). In every case, though, uncompetitive inhibitors have a much larger effect than competitive inhibitors. This is a result of the potentiation of inhibitor binding by substrate. Thus, a rise in substrate concentration can overcome a competitive inhibitor and restore metabolic flux, but this becomes increasingly difficult for an uncompetitive inhibitor if the initial flux (or substrate concentration) is high. Indeed, the uncompetitive inhibitor causes dramatic rises in substrate concentration with potentially equally dramatic consequences for the cell. Lithium illustrates this perfectly. In cortical slice experiments, for example, lithium alone up to high concentrations has little effect on Ins(1)P levels. This is because the starting concentration of substrate is low, and lithium blockade can be ameliorated by a modest increase in Ins(1)P concentration. In the presence of carbachol, however, Ins(1)P is higher and flux through the phosphoinositide cycle is increased. Lithium blockade now leads to a much greater rise in Ins(1)P since it has become much harder to overcome the uncompetitive block. Table 8.2 gives calculations which illustrate this point. For a variety of starting Ins(1)P concentrations the rise in Ins(1)P required to overcome inhibition by Li$^+$ at a concentration equal to its K$_i$ value has been calculated. Thus, only a 2-fold increase is required with a starting Ins(1)P of 0.04 mM. However, when Ins(1)P is 0.12 mM, a 3-fold rise (such as that encountered with carbachol in a cortical slice experiment), a further 13-fold increase in Ins(1)P to 1.56 mM is required to overcome inhibition by lithium. The actual values are extremely sensitive to the flux rate, particularly

TABLE 8.3. Properties of Ins(1,4)P$_2$/Ins(1,3,4)P$_3$ 1-phosphatase.

Molecular	Native M$_r$	44,000–48,000 (13,24,16)
	Subunit M$_r$	40,000–44,000 (13,14,16)
Catalytic	K$_m$ (Ins(1,4)P$_2$)	0.04 mM (13), 0.004 mM (14), 0.03 mM (16)
	K$_m$ (Ins(1,3,4)P$_3$)	0.5 mM (13), 0.021 mM (14)
	V$_{max}$ (Ins(1,4)P$_2$)	61 (13), 50 (14), 19 µmol/min/mg (16)
	V$_{max}$ (Ins(1,3,4)P$_3$)	63 (13), approx 35 µmol/min/mg (14)

Values are for the purified bovine brain (13,14) or rat brain (16) enzymes. V$_{max}$ for Ins(1,3,4)P$_3$ from ref. 14 was quoted as 1.2 to 1.5 times less than V$_{max}$ for Ins(1,4)P$_2$.

when this is close to a v/V$_{max}$ of 0.5 (73) but are similar to values in the literature for cortical slice experiments (39,61,62). Thus, an uncompetitive inhibitor is the equivalent for enzymes of a "use-dependent" antagonist of a receptor. A cell whose phosphoinositide cycle is unstimulated will tolerate lithium with scarcely any modulation of the concentrations of cycle metabolites. A heavily stimulated cell, however, will undergo drastic swings in metabolite levels to maintain its activity in the presence of lithium. This provides a good explanation for the apparent tissue-specificity of lithium. The specificity lies not in any major enzymatic differences between cells but in the degree of stimulation of their phosphoinositide cycle.

It follows from the previous discussion that the therapeutic activity of lithium might depend as much on its mode of inhibition as on its target and this should be considered in any attempts to produce organic alternatives to lithium as novel pharmaceuticals.

Ins(1,4)P2/Ins(1,3,4)P3 1-phosphatase

In principle, the same considerations apply to the enzyme which dephosphorylates Ins(1,4)P$_2$ and Ins(1,3,4)P$_3$ to Ins(4)P and Ins(3,4)P$_2$, respectively. The enzyme has been purified by three groups recently (13,14,16) and there is agreement that it is a monomeric protein of M$_r$ 40,000 (Table 8.3). One group incorrectly identified the product of Ins(1,4)P$_2$ hydrolysis as Ins(1)P (16) and Gee et al. (13) finally established the substrate and product specificity of the homogeneous enzyme. The enzyme is specific for Ins(1,4)P$_2$ and Ins(1,3,4)P$_3$, hydrolyzing no other inositol mono, bis, or polyphosphates. The enzyme requires Mg^{2+} for activity and is uncompetitively inhibited by lithium. There is a considerable disagreement between various groups as to the K$_m$ for the two substrates and the K$_i$ values for lithium. This requires further investigation but is probably a result of the use of different buffers, or it may be a consequence of some unidentified regulatory mechanism. The preparations of Gee et al. (13) and Inhorn and Majerus (14) are similar in that the K$_i$ for lithium towards Ins(1,3,4)P$_3$ hydrolysis is considerably lower than that towards Ins(1,4)P$_2$ hydrolysis, but the K$_m$ for the former substrate is higher.

That lithium can elevate $Ins(1,4)P_2$ and $Ins(1,3,4)P_3$ in tissue slices or cells has been clearly shown, but higher concentrations are required than for elevation of $Ins(1)P$ and the extents of elevation are less dramatic (see earlier). The extent to which elevation of $Ins(1,4)P_2$ and $Ins(1,3,4)P_3$ can be accounted for by direct inhibition of the enzyme is unclear in the absence of precise data on the kinetic constants for the enzyme and the absolute concentrations of the two metabolites. It should be pointed out that $Ins(1,4)P_2$ and $Ins(1,3,4)P_3$ are also mutually competitive since the same enzyme is responsible for their hydrolysis. Majerus (38) has proposed that elevation of $Ins(1,3,4)P_3$ might be a therapeutically important effect of lithium. Inhibition of 1-phosphatase will promote metabolism via $Ins(1,3)P_2$ and $Ins(1,3,4,6)P_4$ with consequences as yet undefined.

Other Phosphatases and Phosphoinositide Cycle Enzymes

The similarity between inositol monophosphatase and $Ins(1,4)P_2/Ins(1,3,4)P_3$ 1-phosphatase in their inhibition by lithium strongly suggests that they are mechanistically and even structurally related, despite the obvious differences in their molecular parameters. However, none of the other phosphatases of the cycle are inhibited by lithium with one curious exception recently identified. In the slime mold, *Dictyostelium discoideum*, $Ins(1,4,5)P_3$ can be degraded by a novel 1-phosphatase to $Ins(4,5)P_2$ and this enzyme is lithium-sensitive, although the mode of inhibition has not been established (74).

No other enzymes of the phosphoinositide cycle are directly affected by lithium, so far is known, although an indirect stimulation of phospholipase C was found by Volonte and Racker (75) in membranes from PC12 cells differentiated with nerve growth factor. The stimulation was suggested to result from increased G protein activation since lithium at only 1.1 mM increased GTP binding to these membranes by up to fivefold (76). These observations are reminiscent of the considerable body of work suggesting that lithium at therapeutic levels inhibits noradrenaline-activated adenylate cyclase, either directly or by inhibition of G protein (Gs) activation. Avissar et al. (77) showed that lithium inhibited the increase in GTP binding to cortical membranes induced not only by isoprenaline (operating via Gs) but also by carbachol (operating via Gi and Go). These effects on receptor-mediated signalling through G proteins remain a plausible candidate for the therapeutic target for lithium. However, the exact site has not been identified and the mechanism does not provide any explanation for the central selectivity of lithium action.

Concluding Discussion

The previous sections have described the effects of lithium on phosphoinositide turnover at various levels of biological complexity. It is now appropriate to return to the two main issues proposed at the beginning of this chapter. The first of these is whether the action of lithium on the cycle can be explained by inhibition of a single step in the pathway. Certainly the most overt effect of lithium, the elevation of Ins(1)P (or other monophosphates) and the lowering of inositol, is clearly a result of monophosphatase inhibition. There are few data on the effects of lithium in vivo on other inositol phosphates in brain but there is nothing to support a substantial effect on Ins(1,4)P$_2$ and Ins(1,3,4)P$_3$ (78,79). However, elevation of these in vitro occurs and is most likely a result of inhibition of the 1-phosphatase. Lithium also gives rise to elevation of CMPPA, and this can be adequately explained by monophosphatase inhibition (63). The observation that elevation of CMPPA has a higher ED$_{50}$ for lithium than elevation of Ins(1)P does not argue against this explanation. CMPPA will rise as a result of limitation of the rate of PI synthesis by the decline in inositol concentration. If the inositol concentration is considerably higher than the Km of the enzyme for inositol, then inositol depletion will have to be severe (i.e., at high lithium) before flux is reduced and CMPPA rises to compensate. Since CMPPA, PA, and DAG are reversibly interconvertible, the elevation by lithium of PA and DAG becomes explicable in terms of an underlying inhibition of monophosphatase. The proposal that lithium can therefore indirectly stimulate protein kinase C opens up all kinds of possibilities for further effects, both acute and chronic.

One would expect that such effects would be prevented by replenishing the inositol pool, and in vitro this is certainly so, since the lithium-induced rise in CMPPA and PA in parotid slices is reversed by inositol (63). The observation that inositol does not restore agonist-induced phosphoinositide turnover in cortical slices from rats pretreated with lithium (56) does not weaken the argument, since the timescale of the adaptive changes may be much greater than that attainable in vitro. It might be more profitable to explore longer term effects of lithium and their reversal by inositol in a cell line which can be grown or maintained in the presence of one or both of these compounds. At our present state of knowledge, therefore, chronic effects of lithium on receptor/G protein/ effector interactions could well be brought about by an underlying monophosphatase inhibition. Acute or in vitro effects are much harder to explain, and perhaps the most interesting of these is the delayed inhibition of carbachol-induced Ins(1,3,4,5)P$_4$ accumulation (39,62). The fact that inhibition occurs at doses of lithium lower than those needed for CMPPA accumulation argues against an effect mediated by DAG. In addition, carbachol-stimulated Ins(1,3,4,5)P$_4$ accumulation is unaffected by phorbol myristate acetate suggesting that indirect stimulation of protein kinase C is not responsible for the action of lithium. However, it is possible that the Ins(1,3,4,5)P$_4$ effect is a result of inositol monophosphate accumulation via an as yet unidentified regulatory mechanism. Thus, the first of the main issues is by no means resolved, but a plausible case can be made that inhibition of the inositol monophosphatase could directly or indirectly be responsible for most of the actions of lithium on the phosphoinositide cycle. The second issue is whether modulation of the phosphoinositide cycle provides a reasonable explanation for the therapeutic action of lithium in manic depression. We are on much weaker ground here, although there are a few observations of some relevance. The original proposal for the antimanic effect of lithium was that blockade of monophosphatase would lead to a depletion of phospholipase C substrates and therefore dampen down phosphoinositide-mediated signal transduction. The theory has been elaborated by Drummond (55) to include both antimanic and antidepressant actions by postulating dampening effects on both inhibitory and excitatory neurones. While the evidence for depletion of phosphoinositides (especially PIP_2) is thin, it is certainly true that lithium causes a decrease in cholinergic-stimulated phosphoinositide turnover. Worley et al. (80) have shown that this can result in altered synaptic transmission in the hippocampus. Adenosine blocks the CA1 population spike elicited by stimulation of Schaffer collaterals and the inhibitory effect of adenosine is itself blocked by muscarinic agonists. In the presence of carbachol the adenosine effect is dose- and time-dependently restored by lithium. Phorbol esters can also antagonize the adenosine effect, showing that protein kinase C activation is involved, and this effect is not blocked by lithium. The interpretation given was that lithium dampens phosphoinositide turnover elicited by carbachol, thereby reducing the extent of protein kinase C activation by DAG. There is a danger here of arguing both ends against the middle. After all, an explanation of the decreased muscarinic response in cortex was that lithium enhances DAG and thereby inhibits muscarinic signal transduction via protein kinase C. The adenosine effect is attributed to decreased DAG production in the presence of lithium. Perhaps one can reconcile these arguments by an acute effect of lithium causing increased DAG leading to a chronic (and perhaps selective) decrease in cholinergic transmission and a chronic decrease in DAG. Alternatively, the decreased cholinergic transmission could be explained by a direct effect of lithium on receptor/G protein coupling. Whatever the underlying detailed explanation, these experiments clearly demonstrate that lithium action on PI turnover can modulate synaptic transmission and this idea can be extended, with some imagination, to explain how lithium exhibits a normalizing effect on both manic and depressive episodes of bipolar disorder.

References

1. Bunney, E., & Garland-Bunney, B.L. (1987). Mechanisms of action of lithium in affective illness: Basic and clinical implications. In H.Y. Meltzer (Ed.), *Psychopharmacology: The third generation of progress* (pp. 553–565). New York: Raven Press.
2. Wood, A.J., & Goodwin, G.M. (1987). A review of the biochemical and neuropharmacological actions of lithium. *Psychol. Med.*, *17*,579–600.
3. Nahorski, S.R., Kendall, D.A., & Batty, I. (1986). Receptors and phosphoinositide metabolism in the central nervous system. *Biochem. Pharmacol.*, *35*,2447–2453.
4. Berridge, M.J. (1986). Cell signalling through phospholipid metabolism. *J. Cell. Sci. Suppl.*, *4*,137–153.
5. Berridge, M.J. (1987). Inositol trisphosphate and diacylglycerol: Two interacting second messengers. *Annu Rev Biochem.*, *56*,159–193.
6. Downes, C.P., Mussat, M.C., & Michell, R.H. (1982). The inositol trisphosphate phosphomonoesterase of the human erythrocyte membrane. *Biochem. J.*, *203*,169–177.
7. Connolly, T.M., Bross, T.E., & Majerus, P.W. (1985). Isolation of a phosphomono-

esterase from human platelets that specifically hydrolyses the 5-phosphate of inositol 1,4,5-trisphosphate. *J. Biol. Chem.*, *260*,7868–7874.

8. Connolly, T.M., Bansal, V.S., Bross, T.E., et al. (1987). The metabolism of tris and tetraphosphates of inositol by 5-phosphomonoesterase and 3-kinase enzymes. *J. Biol. Chem.*, *262*,2146–2149.

9. Hansen, C.A., Johanson, R.A., Williamson, J.R., et al. (1987). Purification and characterization of two types of soluble inositol phosphate 5-phosphomonoesterases from rat brain. *J. Biol. Chem.*, *262*,17319–17326.

10. Balla, T., Baukal, A.J., Guillemette, G., et al. (1986). Angiotensin-stimulated production of inositol trisphosphate isomers and rapid metabolism through inositol 4-monophosphate in adrenal glomerulosa cells. *Proc. Natl. Acad. Sci.* (USA), *83*, 9323–9327.

11. Delvaux, A., Dumont, J.E., & Erneux, C. (1987). The metabolism of inositol-4-monophosphate in rat mammalian tissues. *Biochem. Biophys, Res. Commun.*, *145*, 59–65.

12. Ragan, C.I., Watling, K.J., Gee, N.S., et al. (1988). The dephosphorylation of inositol 1,4-bisphosphate to inositol in liver and brain involves two distinct Li^+-sensitive enzymes and proceeds via inositol 4-phosphate. *Biochem. J.*, *249*,143–148.

13. Gee, N.S., Reid, G.G., Jackson, R.G., et al. (1988). Purification and properties of inositol-1,4-bisphosphatase from bovine brain. *Biochem. J.*, *253*,777–782.

14. Inhorn, R.C., & Majerus, P.W. (1988). Properties of inositol polyphosphate 1-phosphatase. *J. Biol. Chem.*, *263*,14559–14565.

15. Story, D.J., Shears, S.B., Kirk, C.J., et al. (1984). Stepwise enzymatic dephosphorylation of inositol 1,4,5-trisphosphate to inositol in liver. *Nature* (London), *312*, 374–376.

16. Takimoto, K., Motoyama, N., Okada, M., et al. (1987). Purification and properties of inositol-1,4-bisphosphate 4-phosphohydrolase from rat brain. *Biochim. Biophys. Acta*, *929*,327–335.

17. Morris, A.J., Storey, D.J., Downes, C.P., et al. (1988). Dephosphorylation of 1D-*myo*-inositol 1,4-bisphosphate in rat liver. *Biochem. J.*, 655–660.

18. Irvine, R.F., Letcher, A.J., Heslop, J.P., et al. (1986). The inositol tris/tetrakisphosphate pathway-demonstration of Ins(1,4,5)P₃ 3-kinase activity in animal tissues. *Nature* (London), *320*,631–634.

19. Johanson, R.A., Hansen, C.A., & Williamson, J.R. (1988). Purification of D-*myo*-inositol 1,4,5,-trisphosphate 3-kinase from rat brain. *J. Biol. Chem.*, *263*,7465–7471.

20. Batty, I.R., Nahorski, S.R., & Irvine, R.F. (1985). Rapid formation of inositol 1,3,4,5-tetrakisphosphate following muscarinic receptor stimulation of rat cerebral cortical slices. *Biochem. J.*, *232*,211–215.

21. Shears, S.B., Parry, J.B., Tang, E.K.Y., et al. (1987). Metabolism of D-*myo*-inositol 1,3,4,5,-tetrakisphosphate by rat liver, including the synthesis of a novel isomer of *myo*-inositol tetrakisphosphate. *Biochem. J.*, *246*,139–147.

22. Erneux, C., Delvaux, A., Moreau, C., et al. (1987). The dephosphorylation pathway of D-*myo*-inositol 1,3,4,5-tetrakisphosphate in rat brain. *Biochem. J.*, *247*,635–639.

23. Doughney, C., McPherson, M.A., & Dormier, R.L. (1988). Metabolism of inositol 1,3,4,5-tetrakisphosphate by human erythrocyte membranes. *Biochem. J.*, *251*,927–929.

24. Höer, D., Kwiatkowski, A., Seib, C., et al. (1988). Degradation of inositol 1,3,4,5-tetrakisphosphates by porcine brain cytosol yields inositol 1,3,4,-trisphosphate and inositol 1,4,5,-trisphosphate. *Biochem. Biophys. Res. Commun.*, *154*,668–675.

25. Cunha-Melo, J.R., Dean, N.M., Ali, H., et al. (1988). Formation of inositol 1,4,5,-trisphosphate and inositol 1,3,4-trisphosphate from inositol 1,3,4,5-tetrakisphosphate and their pathway of degradation in RBL-2H3 cells. *J. Biol. Chem.*, *263*, 14245–14250.

26. Inhorn, R.C., Bansal, V.S., & Majerus, P.W. (1987). Pathway for inositol 1,3,4,-trisphosphate and 1,4-bisphosphate metabolism. *Proc. Natl. Acad. Sci.* (USA), *84*, 2170–2174.

27. Shears, S.B., Kirk, C.J., & Michell, R.H. (1987). The pathway of *myo*-inositol 1,3,4-trisphosphate dephosphorylation in liver. *Biochem. J.*, *248*,977–980.

28. Bansal, V.S., Inhorn, R.C., & Majerus, P.W. (1987). The metabolism of inositol 1,3,4-trisphosphate to inositol 1,3-bisphosphate. *J. Biol. Chem.*, *262*,9444–9447.

29. Dean, N.M., & Moyer, J.D. (1988). Metabolism of inositol bis-, tris- tetrakis- and pentakis-phosphates in GH₃ cells. *Biochem. J.*, *250*,493–500.

30. Balla, T., Guillemette, G., Baukal, A.J., et al. (1987). Metabolism of inositol 1,3,4-trisphosphate to a new tetrakisphosphate isomer in angiotensin-stimulated adrenal glomerulosa cells. *J. Biol. Chem.*, *262*,9952–9955.

31. Hansen, C.A., vom Dahl, S., Huddell, B., et al. (1988). Characterization of inositol 1,3,4-trisphosphate phosphorylation in rat liver. *FEBS Lett.*, *236*,53–56.

32. Stephens, L.R., Hawkins, P.T., Barker, C.J., et al. (1988). Synthesis of *myo*-inositol 1,3,4,5,6-pentakisphosphate from inositol phosphates generated by receptor activation. *Biochem. J.*, *253*,721–733.

33. Vallejo, M., Jackson, T., Lightman, S., et al. (1987). Occurrence and extracellular actions of inositol pentakis- and hexakisphosphates in mammalian brain. *Nature* (London), *330*,656–658.

34. Stephens, L., Hawkins, P.T., Carter, N., et al. (1988). L-*myo*-inositol 1,4,5,6-tetrakisphosphate is present in both mammalian and avian cells. *Biochem. J.*, *249*,271–282.

35. Morgan, R.O., Chang, J.P., & Catt, K.J. (1987). Novel aspects of gonadotropin-releasing hormone action on inositol polyphosphate metabolism in cultured pituitary gonadotrophs. *J. Biol. Chem.*, *262*,1166–1171.

36. Ackermann, K.E., Gish, B.G., Honchar, M.P., et al. (1987). Evidence that inositol-1-phosphate in brain of lithium-treated rats results mainly from phosphatidylinositol metabolism. *Biochem. J.*, *242*,517–524.

37. Siess, W. (1985). Evidence for the formation of inositol 4-monophosphate in stimulated human platelets. *FEBS Lett.*, *185*,151–156.

38. Majerus, P.W., Connolly, T.M., Bansal, V.J., et al. (1988). Inositol phosphates: Synthesis and degradation. *J. Biol. Chem.*, *263*,3051–3054.

39. Whitworth, P., & Kendall, D.A. (1988). Lithium selectively inhibits muscarinic receptor-stimulated inositol tetrakisphosphate accumulation in mouse cerebral cortex slices. *J. Neurochem.*, *51*,258–265.

40. Dillon, S.B., Murray, J.J., Verghese, M.W., et al. (1987). Regulation of inositol phosphate metabolism in chemoattractant-stimulated human polymorphonuclear leukocytes. *J. Biol. Chem.*, *262*,11546–11552.

41. Dixon, J.F., & Hokin, L.E. (1985). The formation of inositol 1,2-cyclic phosphate on agonist stimulation of phosphoinositide breakdown in mouse pancreatic minilobules. *J. Biol. Chem.*, *260*,16068–16071.

42. Sekar, M.C., Dixon, F.J., & Hokin, L.E. (1987). The formation of inositol 1,2-cyclic 4,5-trisphosphate and inositol 1,2-cyclic 4-bisphosphate on stimulation of mouse pancreatic minilobules with carbamylcholine. *J. Biol. Chem.*, *262*,340–344.

43. Ishii, H., Connolly, T.M., Bross, T.E., et al. (1986). Inositol cyclic trisphosphate [inositol 1,2-(cyclic)-4,5-trisphosphate] is formed upon thrombin stimulation of human platelets. *Proc. Natl. Acad. Sci.* (USA). *83*,6397–6401.

44. Dixon, J.F., & Hokin, L.E. (1987). Inositol 1,2-cyclic 4,5-trisphosphate concentration relative to inositol 1,4,5-trisphosphate in pancreatic minilobules on stimulation with carbamylcholine in the absence of lithium. *J. Biol. Chem.*, *262*,13892–13895.

45. Hughes, A.R., Takemura, H., & Putney, J.W., Jr. (1988). Kinetics of inositol 1,4,5-trisphosphate and inositol cyclic 1:2,4,5-trisphosphate metabolism in intact rat parotid acinar cells. *J. Biol. Chem.*, *263*,10314–10319.

46. Wilson, D.B., Connolly, T.M., Bross, T.E., et al. (1985). Isolation and characterisation of the inositol cyclic phosphate products of polyphosphoinositide cleavage by phospholipase C. Physiological effects in permeabilised platelets and Limulus photoreceptor cells. *J. Biol. Chem.*, *260*,13496–13501.

47. Allison, J.H., & Stewart, M.A. (1971). Reduced brain inositol in lithium-treated rats. *Nature New Biol.*, *233*,267–268.

48. Allison, J.H., Boshans, R.L., Hallcher, L.M., et al. (1980). The effects of lithium on *myo*-inositol levels in layers of frontal cerebral cortex, in cerebellum, and in corpus callosum of the rat. *J. Neurochem.*, *34*,456–458.

49. Sherman, W.R., Gish, B.G., Honchar, M.P., et al. (1986). Effects of lithium on phosphoinositide metabolism *in vivo*. *Fed. Proc.*, *45*,2639–2646.

50. Sherman, W.R., Leavitt, A.L., Honchar, M.P., et al. (1981). Evidence that lithium alters phosphoinositide metabolism: Chronic administration elevates primarily D-*myo*-inositol-1-phosphate in cerebral cortex of the rat. *J. Neurochem.*, *36*,1947–1951.

51. Sherman, W.R., Munsell, L.Y., Gish, B.G., et al. (1985). Effects of systemically administered lithium on phosphoinositide metabolism in rat brain, kidney and testis. *J. Neurochem.*, *44*,798–807.

52. Hallcher, L.M., & Sherman, W.R. (1980). The effects of lithium ion and other agents on the activity of *myo*-inositol-1-phosphatase from bovine brain. *J. Biol. Chem.*, *255*, 10896–10901.

53. Renshaw, P.F., Joseph, N.E., & Leigh, J.S., Jr. (1986). Chronic dietary lithium induces increased levels of myo-inositol-1-phosphatase activity in rat cerebral cortex homogenates. *Brain Res.*, *380*,401–404.

54. Berridge, M.J., Downes, C.P., & Hanley, M.R. (1982). Lithium amplifies the agonist dependent phosphatidylinositol responses in brain and salivary glands. *Biochem. J.*, *206*,587–595.

55. Drummond, A.H. (1987). Lithium and inositol lipid-linked signalling mechanisms. *TIPS*, *8*,129–133.

56. Kendall, D.A., & Nahorski, S.R. (1987). Acute and chronic lithium treatments influence agonist and depolarization-stimulated inositol phospholipid hydrolysis in rat cerebral cortex. *J. Pharm. Exp. Therap.*, *241*,1023–1027.

57. Casebolt, T.L., & Jope, R.S. (1987). Chronic lithium treatment reduces norepinephrine-stimulated inositol phospholipid hydrolysis in rat cortex. *Eur. J. Pharmacol.*, *140*,245–246.

58. Newman, M.E., & Lehrer, B. (1988). Effects of lithium and desipramine-administration on agonist-stimulated inositol phosphate accumulation in rat cerebral cortex. *Biochem. Pharmacol.*, *37*,1991–1995.

59. Drummond, A.H., & Raeburn, C.A. (1984). The interaction of lithium with thyrotropin releasing hormone-stimulated lipid metabolism in GH_3 pituitary tumor cells. *Biochem. J.*, *224*,129–136.

60. Labarca, R., Janowsky, A., Patel, J., et al. (1984). Phorbol esters inhibit agonist induced ^{3}H-inositol-1-phosphate accumulation in rat hippocampal slices. *Biochem. Biophys. Res. Commun.*, *123*,703–309.
61. Batty, I., & Nahorski, S.R. (1985). Differential effects of lithium on muscarinic receptor stimulation of inositol phosphates in rat cerebral cortex. *J. Neurochem.*, *45*, 1514–1521.
62. Batty, I., & Nahorski, S.R. (1987). Lithium inhibits muscarinic-receptor-stimulated inositol tetrakisphosphate accumulation in rat cerebral cortex. *Biochem. J.*, *247*,797–800.
63. Downes, C.P., & Stone, M.A. (1986). Lithium-induced reduction in intracellular inositol supply in cholinergically stimulated parotid gland. *Biochem. J.*, *234*,199–204.
64. Huang, E.M., & Detwiler, T.C. (1986). The effect of lithium on platelet phosphoinositide metabolism. *Biochem. J.*, *236*,895–901.
65. Thomas, A.P., Alexander, J., & Williamson, J.R. (1984). Relationship between inositol polyphosphate production and the increase in cytosolic free Ca^{2+} induced by vasopressin in isolated hepatocytes. *J. Biol. Chem.*, *259*,5574–5584.
66. Hansen, C.A., Mah, S., & Williamson, J.R. (1986). Formation and metabolism of inositol 1,3,4,5-tetrakisphosphate in liver. *J. Biol. Chem.*, *261*,8100–8103.
67. Godfrey, P.P. (1989). Potentiation by lithium of CMP-phosphatidate formation in carbachol stimulated rat cerebral cortical slices and its reversal by *myo*-inositol. *Biochem. J.*, *258*,621–624.
68. Takimoto, K., Okada, M., Matsuda, Y., et al. (1985). Purification and properties of *myo*-inositol-1-phosphatase from rat brain. *J. Biochem.* (Tokyo), *98*,363–370.
69. Gee, N.S., Ragan, C.I., Watling, K.J., et al. (1988). The purification and properties of *myo*-inositol monophosphatase from bovine brain. *Biochem. J.*, *249*,883–889.
70. Attwood, P.V., Ducep, J-B., & Chanal, M-C. (1988). Purification and properties of *myo*-inositol-1-phosphatase from bovine brain. *Biochem. J.*, *253*,387–394.
71. Meek, J.L., Rice, T.J., & Anton, E. (1988). Rapid purification of inositol monophosphate phosphatase from bovine brain. *Biochem. Biophys. Res. Commun.*, *156*,143–148.
72. Shute, J.K., Baker, R., Billington, D.C., et al. (1988) Mechanism of the *myo*-inositol phosphatase reaction. *J. Chem. Soc. Chem. Commun.*, 626–628.
73. Cornish-Bowden, A. Why is uncompetitive inhibition so rare? A possible explanation, with implications for the design of drugs and pesticides. *FEBS Lett.*, *203*,3–6.
74. van Lookeren Campagne, M.M., Erneux, C., van Eijk, R., et al. (1988). Two dephosphorylation pathways of inositol 1,4,5-trisphosphate in homogenates of the cellular slime mold *Dictyostelium discoideum*. *Biochem. J.*, *254*,343–350.
75. Volonte, C., & Racker, E. (1988). Lithium stimulation of membrane-bound phospholipase C from PC12 cells exposed to nerve growth factor. *J. Neurochem.*, *51*, 1163–1168.
76. Volonte C. (1988). Lithium stimulates the binding of GTP to the membrane of PC12 cells cultured with nerve growth factor. *Neurosci. Lett.*, *87*,127–132.
77. Avissar, S., Schreiber, G., Danon, A., et al. (1988). Lithium inhibits adrenergic and cholinergic increases in GTP binding in rat cortex. *Nature* (London), *331*,441–443.
78. Meek, J.L. (1986). Inositol bis-, tris-, and tetrakis(phosphate)s: Analysis in tissues by HPLC. *Proc. Natl. Acad. Sci.* (USA), *83*,4162–4166.
79. Sun, G.Y., & Huang, SF-L. (1987). Labelling of phosphoinositides in rat brain membranes: An assessment of changes due to post-decapitative ischemic treatment. *Neurochem. Int.*, *10*,361–369.
80. Worley, P.F., Heller, W.A., & Snyder, S.H. (1988). Lithium blocks a phosphoinositide-mediated cholinergic response in hippocampal slices. *Science*, *239*,1428–1429.

9
Effects of Lithium on Cell Growth

Vincent S. Gallicchio

One of the major dominant themes in cell biology is elucidating at the molecular level the mechanisms by which alterations in the external microenvironment influence the proliferation of mammalian cells. It is a well-described phenomenon that a number of agents or conditions markedly influence the rate of cellular proliferation. Hormones, serum or growth factors, when added to cultures of quiescent cells in a short period of time, all can induce a number of metabolic changes leading to the initiation of DNA synthesis and cell division (1–4). These substances are thought to act via their interaction with the cell membrane by the activation of a second signal that is transmitted into the interior of these cells (5–7). In the absence of these factors mammalian cells stop proliferating and enter a phase referred to as the resting phase of Gl/Go, only to assume proliferation when the absent agent is resupplied.

It is also a well-described phenomenon that the cell membrane plays a very important role in these activation processes because alterations or changes in its permeability appear to be a critical first step in the initiation of the events leading to cell proliferation. With altered changes in cell permeability being important for the initiation of cell growth, this process has demonstrated that substances capable of rapid transport like monovalent or divalent cations, due to changes in cell permeability, can influence and therefore may play a significant role as possible intracellular regulators in the maintenance or altered rate of cell proliferation (8–10). The basic structure of biological membranes involves a phospholipid bilayer whereby its very composition provides an environment that allows for the interaction between these inorganic ions and the membrane. This system influences the interactions between growth factors and the membrane to cause changes in cell growth. Since the major constituent of the membrane lipid bilayer is phospholipids, ions therefore can alter cell growth by their ability to transverse this lipid bilayer. The subject of lithium and its ability to interact with the components of the lipid bilayer was extensively detailed in the previous section. The major emphasis of the following section will focus on the ability of lithium to interact with growth factors in modulating cell proliferation.

Lithium, a monovalent cation, has been identified with a number of physiological, biochemical, and biological effects on mammalian tissues in both organ and primary cell cultures (11, 12). Studies have demonstrated a direct action of lithium on the initiation of DNA synthesis while others have indicated lithium promotes the growth of cells only in the presence of growth factors (13). An explanation for this difference in responsiveness has been attributed to the origin of the particular cell type being analyzed. For example, cells of mesenchymal versus epithelium origin may differ due to their phenotypic expression. Lithium effectively initiates DNA synthesis in a concentration-dependent manner in several fibroblastic cell lines even in cell lines nonresponsive to epidermal growth factor, EGF (14). However, the effects of lithium on continuous cell growth needs further analysis. Lithium stimulation may be additive when combined with maximum concentrations of insulin or EGF, suggesting lithium and these growth factors initiate DNA synthesis by different mechanisms. Also, lithium has been demonstrated to react in a similar fashion when combined with mitogens. Lithium, at an inactive concentration when combined with either insulin or EGF, results in a synergistic enhancement of DNA synthesis. This indicates lithium may react similar to that of a mitogen to promote increased DNA synthesis in the presence of what otherwise is an optimum concentration of growth factor (15). Not all fibroblastic cells respond in a similar fashion (12), indicating that certain cells may lack the physiological conditions necessary for lithium to react as a mitogen in response to growth factor stimulation.

Under these growth conditions, ions like lithium may induce their activity as the result of alterations or changes in membrane ion flux. In some cell systems, serum and hormone stimulated growth effects have been attributed to alterations in monovalent cation changes induced by mitogens (16,17). Because lithium is partially transported into cells by mechanisms that are identical to those that are responsible for the transport of sodium, it has been suggested that the mitogenic effects of lithium are partially due to this increased intracellular ion concentration. However, growth effects that are attributable to increased sodium transport are usually dependent upon: 1) high cell density, and 2) presence of serum, conditions that are not dependent for lithium-induced growth stimulation (18). Both magnesium and calcium have been proposed to be involved in the processes that regulate cell proliferation and have demonstrated that the uptake of sodium may influence the transport of these divalent cations (16). By using sodium transport mechanisms, lithium has also been implicated as capable both to influence and regulate the intracellular concentrations of divalent cations. In fact, lithium has been demonstrated to enhance calcium uptake in isolated fat cells and block its uptake in the endoplasmic reticulum (19,20).

Cell growth and regulation by lithium have implicated cAMP especially with respect to the role played by calcium in the activation of adenylate cyclase activity. In certain systems, this pathway of cAMP activation can be stimulatory for certain cell systems and inhibitory for others (21,22). Under these conditions, lithium can inhibit cell growth where cAMP is stimulatory and promote cell growth where cAMP is inhibitory for cell growth, therefore lithium may regulate cell proliferation by its ability to influence the activity of adenylate cyclase.

Lithium has been shown to be an effective stimulator of glucose transport in certain cell types (23,24). This activity of lithium indicates that this monovalent cation has a similar biological action when compared to insulin whether the action is in vivo or in vitro. This insulin-like activity of lithium has also implied the action of insulin may involve the alteration of cell surface membrane properties that results in changes in the intracellular concentration of certain regulatory factors. This activity, as indicated, in this insulin-dependent system demonstrates that agents like lithium can alter the processes of certain growth factors to enhance their biological actions (25).

Since the action of many biological functions involves the membrane bilayer, whether it specifically relates to alterations in transport processes or receptor-mediated interactions, the membrane phospholipid composition may be critical to the activity of lithium. This may then directly involve the action of lithium on arachidonic metabolism and the synthesis of prostaglandins. Lithium, within therapeutic concentrations (2 mM), has been shown to be an effective agent in the inhibition of PGE1 synthesis. This reported activity is important especially with respect to other biological effects attributed to lithium, for example 1) platelets from patients with manic-depressive illness produce greater amounts of PGE1 than normal, suggesting PGE1 may be important in determining the cyclic mood patterns of affected individuals and 2) the interpretations of the hematopoietic effects of lithium (see Chapter 6 by Gallicchio elsewhere in this text). At higher lithium concentrations, there appears to be an inhibition of the release of free fatty acids from membrane phospholipids such that many of the PGs are inhibited (26). These antiprostaglandin effects may explain the anti-inflammatory reactions attributed to lithium. Further experimentation is required in order to more clearly elucidate the relationships between the biological action of lithium and fatty acid metabolism.

References

1. Holley, R.W., & Kiernan, J.A. (1968). "Contact inhibition" of cell division in 3T3 cells. *PNAS*, USA, *60*, 300.
2. Hershko, A., Mamont, P., Schields, R., Tomkins, G.M. (1971). *Nature* (London), *232*, 206.
3. Temin, H.M., Pierson, R.W., Jr., & Dulak, V.C. (1972). In G.H. Rothblat & V.J. Cristofalo (Eds.), *Growth, nutrition and metabolism of cells in culture* (pp. 50–81). New York: Academic Press.
4. Rudland, P.S., Seifert, W., & Gospodarowicz, D. (1974). Growth control in cultured mouse fibroblasts: Induction of the pleiotypic and mitogenic responses by a purified growth factor. *PNAS*, USA, *71*, 2600.
5. Pardee, A.B., Jimenez de Asua, L., & Rosengurt, E. (1974). In B. Clarkson & R. Baserga (Eds.), *Control of proliferation in animal cells* (pp. 547–561). Cold Spring Harbor, NY: Cold Harbor Laboratory.
6. Rubin, H.A. (1978). Do viruses use calcium ions to shut off host cell functions? *Nature* (London), *271*, 186.

7. McKeehan, W.L., & Ham, R.G. (1978). Calcium and magnesium ions and the regulation of multiplication in normal and transformed cells. *Nature* (London), *275*, 756.

8. Sanui, H., & Rubin, H.A. (1978). Membrane bound and cellular cationic changes asociated with insulin stimulation of cultured cells. *J. Cell Physiol.*, *96*, 265.

9. Dulbecco, R., & Elkington, J. (1975). Induction of growth in resting fibroblastic cell cultures by Ca. *J. PNAS, USA*, *72*, 1584.

10. Mamont, P.S., Bohlen, T., Milann, P., Bey, F., Schuber, F., & Tardif, C. (1976). x-Methyl ornithine, a potent competitive inhibitor .pa of ornithine decarboxylase, blocks proliferation of rat hepatoma cells in culture. *PNAS, USA*, *73*, 1626.

11. Hori, C., & Oka, T. (1979). Induction of lithium ion on multiplication of mouse mammary epithelium in culture. *PNAS, USA*, *76*, 2823.

12. Ptashne, K., Stockdale, F., & Conlon, S. (1980). Initiation of DNA synthesis in mammary epithelium and mammary tumors by lithium ions. *J. Cell Physiol.*, *103*, 41.

13. Hart, D. (1979). Potentiation of phytohemagglutinin stimulation of lymphoid cells by lithium. *Exp. Cell Res.*, *119*, 47.

14. Pruss, R., & Herschman, H. (1979). Variants of 3T3 cells lacking mitogenic response to epidermal growth factor. *PNAS, USA*, *74*, 3918.

15. Smith, J., & Rosengurt, E.J. (1978). Lithium transport by fibroblastic mouse cells: Characterization and stimulation by serum and growth factors in quiescent cultures. *J. Cell Physiol.*, *97*, 441.

16. Rozengurt, E., & Mendoza, S. (1980). Monovalent ion fluxes and the control of cell proliferation in cultured fibroblasts. *Ann. NY Acad. Sci.*, *339*, 175.

17. Frantz, C., Nathan, D., & Scher, C. (1981). Intracellular univalent cations and the regulation of the BALB 1c-3T3 cell cycle. *J. Cell Biol.*, *88*, 51.

18. Toback, F. (1980). Induction of growth in kidney epithelial cells in culture by Na^+. *PNAS, USA*, *77*, 6654.

19. Clausen, T., Elbrink, J., & Martin, B. (1974). Insulin controlling calcium distribution in muscle and fat cells. *Acta Endocrinol* 77, suppl., *191*, 137.

20. DeMeis, L. (1971). Alosteric inhibition by alkali ions of the Ca^{2+} uptake and adenosine triphosphatase activity of skeletal muscle microsomes. *J. Biol. Chem.*, *246*, 4764.

21. Gelfand, E., Dosch, H., Hastings, D., Shore, A. (1979). Lithium: A Modulator of Cyclic AMP Dependent Events in Lymphocytes? Science *203*, 365.

22. Dousa, T., & Hechter, O. (1970). The effect of NaCl & LiCl on vasopressin-sensitive adenyl cyclase. *Life Sci.*, *9*, 765.

23. Haugaard, E.S., Frazer, A., Mendels, J., & Haugaard, N. (1975). Metabolic and electrolyte changes produced by lithium ions in the isolated rat diaphragm. *Biochem. Pharmacol.*, *24*, 1187.

24. Mickel, R.A., Hallidy, L., Haugaard, N., & Haugaard, E.S. (1978). Stimulation by lithium ions of the incorporation of [u−^{14}C] glucose into glycogen in rat brain slices. *Biochem. Pharmacol.*, *27*, 799.

25. Ryback, S.M., & Stockdale, F.E. (1981). Growth effects of lithium chloride in BALB/c-3T3 fibroblasts and Madin-Darby canine kidney epithelial cells. *Exper. Cell Res.*, *136*, 263.

26. Manku, M.S., Horrobin, D.F., & Karmazyn, M. (1979). Prolactin and zinc effects on rat vascular reactivity: Possible relationship to dihomo-γ-linolenic acid and to prostaglandin synthesis. *Endocrinology*, *104*, 774.

10
The Interaction of Lithium with Magnesium-Dependent Enzymes

ARNE GEISLER AND ARNE MØRK

This overview on the influence of lithium (Li^+) upon enzymes regulated by magnesium (Mg^{2+}) is not intended to be comprehensive but mainly to concentrate on some enzymes relevant to neuronal processes. The interaction between Li^+ and Mg^{2+} is of interest since these two ions share a physiochemical similarity, and this is why it has been proposed that Li^+ exerts its mood-stabilizing effects or its toxic symptoms by interfering with Mg^{2+}-regulated processes. Thus, Lazarus (37) notes that Li^+ has been shown to interfere with various enzymes, for example, adenylate cyclase, Mg^{2+}-ATPase, cholinesterase, DNA polymerase, pyruvate kinase, tyrosine aminotransferase, and tryptophan hydroxylase. In addition, some observations on Mg^{2+} metabolism in manic-depressive patients will be mentioned.

Clinical Observations

It has been observed that urine-Mg^{2+} is increased in manic-depressive patients on Li^+ treatment but not in untreated patients (46). Further, Alexander et al. (2) and Glen (25) have observed that Mg^{2+}-ATPase activity is increased in patients chronically treated with Li^+. In a study by Carman and co-workers (16) it was found that patients having a high plasma Ca^{2+}/Mg^{2+} ratio (> 2.62) responded to Li^+ to a higher degree than patients having a lower ratio. In addition, it was also observed that nearly all patients in whom the level of serum-Mg^{2+} rose in the first 5 days of treatment improved, while a poor treatment response occurred in patients with a decreased serum-Mg^{2+}. Such observations have supported the hypothesis that the therapeutic effects of Li^+ may in part be related to the Mg^{2+} metabolism.

Some authors have also observed alterations in plasma-Mg^{2+} during abnormal mood changes (24,29,57).

In this overview it is our intention to review a part of the basis for these assumptions.

The Physiochemical Properties of Lithium and Magnesium

Lithium is a monovalent cation which only occurs in extremely low concentrations in the human body. Lithium can not be considered in any sense to be an obligatory trace metal for the human organism in contrast to such ions as selenium, vanadium, manganese, and zinc (19). However, Li^+ does occur in our environment in low concentrations, e.g., seawater contains about 0.03 mM (44). During conventional Li^+ therapy a concentration of Li^+ about 0.7 mmol/L in the plasma is aimed for. The distribution of Li^+ between the extracellular and intracellular space is roughly identical; this presumably also holds for nerve cells.

Lithium belongs to group 1A of the periodic system. Its ionic radius is 0.60 A and nearly identical to that of Mg^{2+} which has an ionic radius of 0.65 A. However, the corrected hydrated radius of Li^+ is 3.40 A, which is nearly similar to that of calcium (Ca^{2+}) having a hydrated radius of 3.21 A, while this property of Mg^{2+} is 4.65 A (9). This may indicate that Li^+ can both interfere with Mg^{2+}- and Ca^{2+}-dependent processes. Both Li^+ and Mg^{2+} have some tendency to form coordination bonds with nitrogen and they may therefore both possess a possibility to stick to protein molecules, altering their charges and consequently their activities. In accordance, Fossel et al. (23) have shown that Li^+ in vitro displaces Mg^{2+} and Ca^{2+} from their membrane binding sites.

Magnesium is the fourth most abundant cation in the body and the second most abundant intracellular ion, being exceeded only by potassium. In contrast to Li^+, Mg^{2+} has been assumed to be of significance for several cell functions including neuronal processes.

Psychiatric Disturbances Caused by Hypomagnesemia

Magnesium deficiency is not only seen when induced in experimental animals but may occur in humans in various neuropsychiatric conditions.

Thus, Durlach (20) has observed that Mg^{2+} depletion can provoke tetany combined with neurotic symptoms, which responded to Mg^{2+} therapy. Similarly, Seelig et al. (67) observed latent tetany and anxiety in a patient with a marginal deficiency of Mg^{2+}. Wacker and Parisi (72) have reported that psychiatric disturbances such as depression and psychotic behavior sometimes can occur in patients with severe Mg^{2+} deficiency. Hall and Jaffe (27) have observed an organic brain syndrome associated with Mg^{2+} deficiency where symptoms of dementia disappeared when the Mg^{2+} concentration was normalized. Matzen and Martin (45) described a female patient who became disorientated, paranoid, anxious, and affect-labile due to Mg^{2+} deficiency induced by cancer chemotherapy which included cisplatinum. Further, Durlach (21) has recently summarized neurological and psychic symptoms such as anxiety, insomnia, and hyperventilation due to hypomagnesemia.

Furthermore, Christiansen et al. (18) found that the concentration of bone minerals (Ca^{2+} and Mg^{2+}) were lowered in the forearm in manic-depressive patients treated with Li⁺ compared with the control group. In contrast the serum levels of Ca^{2+} and Mg^{2+} were increased in some patients, 12% and 30%, respectively. Increase in the serum level of Mg^{2+} can be induced in experimental animals treated chronically with Li⁺; this may support the assumption that the hypermagnesemia in Li⁺-treated patients is secondary to the given treatment.

The role of Mg^{2+} in manic-depressive disorders has only been studied to a limited degree and the observed changes in Mg^{2+} metabolism have been somewhat contradictory.

Ramsey et al. (63) reported that patients classified as having primary affective disorders had normal Mg^{2+} erythrocyte level. Similarly, Naylor et al. (57) reported that serum Mg^{2+} levels were normal in manic-depressive patients while an elevated serum concentration was found by Cade (14) and Bjørum (12). However, Frizel et al. (24) observed a decrease in serum Mg^{2+} during depression. There has been observed a relationship between a low concentration of Mg^{2+} in CSF and suicide attempts. However, some studies have shown that the Mg^{2+} concentration in CSF in depressed patients was similar to that of controls and that the concentration of Mg^{2+} did not change upon recovery (6,13).

Lithium has been shown in most studies to enhance the level of Mg^{2+} in serum (3,18,47,59). The mechanism by which Li⁺ enhances serum Mg^{2+} is unknown but may be of interest since this effect can be specific for Li⁺.

On the other hand, several studies have suggested a relationship between Ca^{2+} metabolism and bipolar affective disorder (15). Therefore, when relating the significance of Mg^{2+} on mood disturbances it is appropriate to also consider the levels of intracellular Ca^{2+}. At present it is unknown whether there is an alteration in intracellular Mg^{2+} in neurons in manic and/or depressed patients.

Role of Magnesium in the Function of Brain Enzymes

The first observation of the activating effect of Mg^{2+} upon enzyme activity was done by Erdtman (22) who observed that Mg^{2+} increases mammalian renal alkaline phosphatase activity. A review of the significance of Mg^{2+} for the activities of various enzymes has been given by Wacker (73). Several Mg^{2+}-dependent enzymes influence the function of central neurons and may be relevant for the biological basis of mood regulation.

Adenylate Cyclase

Several studies have demonstrated that Li⁺ in vitro and ex vivo, that is, after chronic treatment, inhibits stimulated adenylate cyclase activities in several brain regions. This may be of interest because other treatments of mood disturbances, for example, treatment with antidepressants (polycyclic antidepressants and mono amine oxidase inhibitors) and ECT also modify the activity of the adenylate cyclase, especially the enzyme linked to beta-adrenoceptors (70). In addition, this

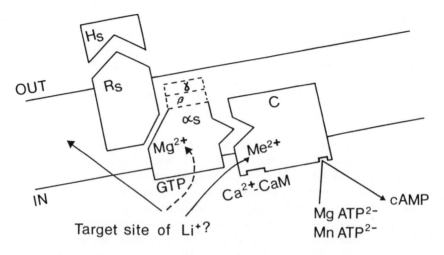

Me^{2+}: Stimulatory : Mg^{2+}, Mn^{2+}
   Inhibitory : Ca^{2+}

FIGURE 10.1. Schematic outline of the basic components of the stimulatory adenylate cyclase system indicating potential sites of action of lithium. The binding of a hormone (H_s) to the stimulatory receptor (R_s) leads to an interaction of H_s-R_s with the GTP-binding protein ($\alpha_s\beta\gamma$). This GTP- and Mg^{2+}-dependent activation of the GTP-binding protein results in a dissociation of $\alpha_s\beta\gamma$ into α_s and $\beta\gamma$. The active conformation of α_s stimulates the activity of the catalytic unit (C). Me^{2+}: a site on the catalytic unit sensitive to divalent cations. Modified from Mørk & Geisler, 1987b, with permission.

enzyme may be of interest in this connection because adenylate cyclase activity is dependent on Mg^{2+} in a dose-dependent manner (17,69).

Our own studies have shown that the inhibitory effect of Li^+ in vitro on adenylate cyclase is antagonized by Mg^{2+} (49,50). Similar observations have been reported by Newman and Belmaker (58). It is important to note that Mg^{2+} affects the activity of adenylate cyclase by an interaction with two distinct ion sites located at the catalytic subunit (41) and the GTP-binding protein (17). In one of our studies we have demonstrated that Li^+ in vitro is able to interact with the Mg^{2+} site at the catalytic unit (50). In fact, this site may be a divalent cation site sensitive not only to Mg^{2+} but also Ca^{2+}. However, an influence of Li^+ on the GTP-binding protein is also possible. Thus, it has been observed that Li^+ ex vivo reduces hormone-induced GTP-binding (4) and hormone-induced GTP stimulation of adenylate cyclase in rat cortical membranes (51). As far as we know no studies have been performed to elucidate a possible interaction between Li^+ and Mg^{2+} on isolated GTP-binding protein(s). The previously mentioned effects of Li^+ in vitro may be relevant for its therapeutic actions because the in vitro effects of Li^+ could be assumed to occur in vivo as an in situ effect. This assumption does not exclude that Li^+ after chronic treatment may exert other effects which directly or indirectly may interfere with the activity of the adenylate cyclase. In fact we have observed that chronic Li^+ treatment reduces the activity of the enzyme in a Mg^{2+}-independent manner (52) (see Figure 10.1).

ATPases

Na^+/K^+-ATPase

This enzyme is distributed in both excitable and nonexcitable tissues. In brain Na^+/K^+-ATPase is present both in neuronal and glial cells. The main function of the enzyme is to maintain the distribution of these monovalent cations across the cell membrane. It has been proposed that the concentration gradient of K^+ and especially that of Na^+ is altered during affective disorder. In fact, Naylor and Smith (55) have suggested that a genetically based dysfunction of the Na^+ pump activity in affective illness causes increased levels of intracellular Na^+. Other studies, using various cell types and fluids, have suggested an increase of intracellular Na^+ (32). In accordance some studies have shown that Li^+ therapy causes a transient increase in Na^+ excretion (5). However, other studies have not confirmed these observations (31). Clearly, such studies are difficult to interpret due to variations in the methodologies used, the type of depression, and most important, the uncertainty whether the observed changes are caused by Li^+ or by different causes of amelioration of the mood disturbance.

On the other hand, some studies have shown that Li^+ treatment significantly increases the Na^+/K^+-ATPase activity in erythrocyte membranes from patients with recurrent manic-depressive illness (34,54) and in patients with a variety of psychiatric diagnoses, including affective states (30). Interestingly, Alexander et al. (2) compared the activity of Na^+/K^+-ATPase in euthymic patients and healthy controls and noted that Li^+ therapy enhanced the activity of the enzyme in a dose-dependent manner.

The mechanism by which Li^+ stimulates Na^+/K^+-ATPase is unclear but it is known that vanadate is a potent inhibitor of the Na^+/K^+-ATPase. This effect of vanadate is reversed by Li^+ (56). This effect could not be confirmed by MacDonald et al. (40).

Alexander et al. (2) suggest that Li^+ also affects the pituitary causing altered cortisol or other pituitary hormone levels which may influence the function of the erythrocyte membrane. It is therefore possible that Li^+ may alter other cell membranes by a similar mechanism. At present it is unknown whether Li^+ interacts with Mg^{2+} when enhancing the activity of the Na^+/K^+-ATPase. However, this mechanism has been proposed by Gupta and Crollini (26).

Ca^{2+}-ATPase

In nervous tissue Ca^{2+}-ATPase participates in the regulation of the intracellular concentration of Ca^{2+}. This enzyme has been studied in brain slices and synaptosomes where it has been shown that the efflux of Ca^{2+} is reduced when external Na^+ is replaced by Li^+ (71). In contrast, Reading and Isbir (64) found that Li^+ treatment enhanced the activity of Ca^{2+}-ATPase activity in rat cortex. These authors suggest that chronic Li^+ treatment could bring about a reduced level of free intracellular Ca^{2+}. Furthermore, Koenig and Jope (36) found that Li^+ in vitro had no influence on K^+-induced Ca^{2+} influx in rat synaptosomes but chronic Li^+ treatment was observed to increase this influx. In addition, it can be mentioned that Alexander et al. (2) only found a trend toward higher Ca^{2+}-ATPase activity in patients with high plasma Li^+ compared to patients with low plasma Li^+ but this difference did not reach statistical significance. Finally, Linnoila et al. (39) found a higher Ca^{2+}-ATPase activity in patients compared to controls.

Mg^{2+}-ATPase

Sengupta et al. (68) found that erythrocyte membrane Mg^{2+}-ATPase activity was increased in both depressive and manic-depressive patients in comparison with controls. These authors found in accordance with Gupta and Crollini (26) that Li^+ in vitro did not affect the Mg^{2+}-ATPase activity. Alexander et al. (2) however observed a positive correlation between Mg^{2+}-ATPase activity and higher Li^+ levels in plasma (0.85–1.2 mM) in contrast to patients with lower levels (0.33–0.85 mM), but the difference was not significant.

Other Magnesium-dependent Enzymes

Pyruvate Kinase

Pyruvate kinase has been used as a model system for studying the influence of Li^+ on Mg^{2+}-dependent enzymes by Birch and co-workers (10). It is unclear whether this enzyme is of significance for the neurobiology of manic-depressive psychosis. The study by Birch et al. (10) demonstrated that Li^+ inhibited the pyruvate kinase noncompetitively with respect to Mg^{2+}. Thus, this study shows that Li^+ may affect enzymatic activity without interfering with the binding of Mg^{2+}. Additionally, Li^+ did not reduce the activity of the enzyme in the presence of manganese (Mn^{2+}).

Myo-inositol Phosphatase

The activity of this enzyme is also dependent on Mg^{2+}. It was first reported by Berridge et al. (7) that Li^+ in vitro (0.1 mM) reduces the activity of myo-inositol-1-phosphatase and this was subsequently confirmed by several other researchers. This pharmacological effect of Li^+ has matured the hypothesis that Li^+ may exert its therapeutic actions by interfering with the inositol lipid signal pathway (8).

It has also been shown that Li^+ in vitro inhibits the myo-inositol-1-phosphatase noncompetitively with respect to Mg^{2+} (28). In accordance Ragan et al. (62) have reported that the purified cytosolic inositol (1,4) bisphosphate phosphatase has similarities to the myo-inositol-1-phosphatase; both enzymes are inhibited uncompetitively by low concentrations (0.1 mM Li^+) in the substrate.

Tryptophan Hydroxylase

Tryptophan hydroxylase has been detected in almost all tissues including brain. In cerebrum this enzyme is localized in serotonergic neurons and its presence has been used as a marker for these neurons.

Knapp and Mandell (35) have reported that in rats treated for 10 to 21 days with Li^+, the cell body tryptophan hydroxylase was significantly lower compared with controls. The same authors (43) showed that Li^+ in vitro changed the kinetic of tryptophan hydroxylase from a hyperbolic saturation curve to a sigmoid curve. The authors speculated that this change was due to an uncoupling and randomizing process on a fundamental molecular level and that this might be a potential therapeutic action of Li^+. Furthermore, Perez-Cruet et al. (61) and Schubert (65) found that chronic Li^+ treatment increased serotonin formation; however they also observed a concomitant increase of tryptophan concentration in brain suggesting an influence of Li^+ on tryptophan uptake. Such an effect of Li^+ has also been demonstrated by Schwann et al. (66). The activity of brain tryptophan hydroxylase is regulated by a Ca^{2+}-dependent ATP-Mg^{2+} activation (42). This complex influence of Li^+ upon serotonin synthesis may therefore be due to an interaction with Mg^{2+} and/or Ca^{2+} sites.

DNA Polymerase

Magnesium plays several important roles in protein synthesis. Thus, DNA polymerase requires Mg^{2+} for activity (38) and Mg^{2+} is also necessary for the binding of mRNA to ribosomes (48). Studies on the influence of Li^+ on protein synthesis have used high and toxic concentrations of Li^+. It is unknown whether these effects of Li^+ are related to Mg^{2+}. Possibly the effect of Li^+, at least at these concentrations, may be found to be due to several distinct mechanisms (33).

Summary

The studies mentioned in this chapter suggest that Li^+ influences Mg^{2+} metabolism and/or Mg^{2+}-dependent physiological effects on the cellular level in animals as well as in humans. We have especially focused upon certain enzymes. Due to the fact that many enzymes are dependent on Mg^{2+}, several processes could be influenced by Li^+. However, it is not likely that Li^+ possesses an identical affinity to every Mg^{2+} site in the organism because other cations may also affect enzyme activity. For example, concerning some of the previously mentioned enzymes, adenylate cyclase, pyruvate kinase, and myo-inositol-1-phosphatase, Ca^{2+} competitively inhibits adenylate cyclase and pyruvate kinase activities with respect to Mg^{2+} and Mn^{2+} (11,60) but both Ca^{2+} and Mn^{2+} are competitive inhibitors of myo-inositol-1-phosphatase with respect to Mg^{2+} (28). Furthermore, Cech and Maguire (17) have reported that Mn^{2+} is able to distinguish between the Mg^{2+} sites on the GTP-binding protein and the catalytic subunit of the adenylate cyclase in lymphoma cells. Some of the studies have demonstrated that Li^+ competitively inhibits the activity of adenylate cyclase with respect to Mg^{2+} (49,50,58) but inhibits pyruvate kinase (11) and myo-inositol-1-phosphatase (28) noncompetitively with respect to Mg^{2+}. In addition, Li^+ in vitro and ex vivo inhibits Mn^{2+}-stimulated adenylate cyclase (49,58) but no effect of Li^+ is observed when activating pyruvate kinase with Mn^{2+} (11). These observations indicate that such cations may affect enzyme activities differently when interacting with their binding sites. The degree of effect of Li^+ may therefore depend on the nature and character of the cation site. It appears likely that Li^+ does not exert these changes by a single effect. Some of these changes may be direct while others may be indirect. For example, it is probable that Li^+ also modifies Ca^{2+} regulation of various enzymes.

At present, it is obvious that Li^+ does modulate some Mg^{2+}-dependent enzymes but its relevance for the therapeutic profile of Li^+ is unclear and needs further investigation. We have demonstrated that Li^+ in vitro and ex vivo reduces the stimulated adenylate cyclase activity by two distinct mechanisms. This observation is of interest since other antidepressive modalities also inhibit this enzyme.

However, it is not known whether the previously mentioned effects of Li^+ are related to: the antimanic, the antidepressant, the prophylactic, or the toxic effects.

References

1. Aikawa, J.K. (1963). *The relationship of magnesium to disease in domestic animals and in humans.* Springfield, IL: C.C. Thomas.
2. Alexander, D.R., Deeb, M., Bitar, F., & Antum, F. (1986). Sodium-potassium, magnesium, and calcium ATPase activities in erythrocytes membranes from manic-depressive patients responding to lithium. *Biol. Psychiatry, 21,* 997–1007.
3. Aronoff, M.S., Evens, R.G., & Durell, J. (1971). Effect of lithium salts on electrolytes metabolism. *J. Psychiat. Res., 8,* 139–159.
4. Avissar, S., Schreiber, G., Dennon, A., & Belmaker, R.H. (1988). Lithium inhibits adrenergic and cholinergic increases in GTP binding in rat cortex. *Nature, 31,* 440–442.
5. Baer, L., Platman, S.R., & Fieve, R.R. (1970). The role of electrolytes in affective disorders. *Arch. Gen. Psychiat., 22,* 108–113.
6. Bech, P., Kirkegaard C., Bock, E., Johannesen, M., & Rafaelsen, O.J. (1978). Hormones, electrolytes and cerebrospinal fluid proteins in manic-melancholic patients. *Neuropsychobiology, 4,* 99–112.
7. Berridge, M.J., Downes, C.P., & Hanley, M.R. (1982). Lithium amplifies agonist-dependent phosphatidylinositol responses in brain and salivary glands. *Biochem. J., 206,* 587–595.
8. Berridge, M.J. (1986). Inositol triphosphate: A new second messenger. In J.S. Schou, A. Geisler, & S. Norn (Eds.), *Drug receptors and dynamic processes in cells* (pp. 90–104). Copenhagen: Munksgaard.

9. Birch, N.J. (1974). Lithium and magnesium-dependent enzymes. *Lancet*, *2*, 965–966.
10. Birch, N.J., Hughes, M.S., Thomas, G.M.H., & Partridges, S. (1986). Lithium and magnesium: Inorganic pharmacology. *Magnesium-Bull.*, *8*, 145–147.
11. Birch, N.J. (1987). Magnesium and psychiatry: Biochemistry and inorganic pharmacology related to lithium. In B.M. Altura, J. Durlach, & M.S. Seeling (Eds.), *Magnesium in cellular processes and medicine* (pp. 212–218). Basel: Karger.
12. Bjørum, N. (1972). Electrolytes in blood in endogenous depression. *Acta Psychiat. Scand.*, *48*, 59–68.
13. Bjørum, N., Plenge, P., & Rafaelsen O.J. (1972). Electrolytes in cerebrospinal fluid in endogenous depression. *Acta Psychiat. Scand.*, *48*, 533–539.
14. Cade, J.F.J. (1964). A significant elevation of plasma magnesium levels in schizophrenia and depressive states. *Med. J. Aust.*, *1*, 195–196.
15. Carman, J.S., Wayatt, E.S., Smith, W., Post, R.M., & Ballenger, J.C. (1984). Calcium and calcitonin in bipolar affective disorder. In R.M. Post & R.M. Ballenger (Eds.), *Neurobiology of mood disorders* (pp. 340–355). Baltimore/London: Williams & Wilkins.
16. Carman, J.S., Post, R.M., Teplitz, T.A., & Goodwin, F.K. (1974). Divalent cations in predicting antidepressant response to lithium. *Lancet*, *2*, 1454.
17. Cech, S.Y., & Maguire, M.E. (1982). Magnesium regulation of beta-receptor-adenylate cyclase complex. I. Effects of manganese on receptor binding and cyclase activation. *Mol. Pharmacol.*, *22*, 267–272.
18. Christiansen, C., Baastrup, P.C., & Transbøl, I. (1975). Osteopenia and dysregulation of divalent cations in lithium-treated patients. *Neuropsychobiology*, *1*, 344–354.
19. Crammer, J.L. (1985). The problem of lithium mechanisms. In S. Gabay, J. Harris, & B.T. Ho (Eds.), *Metal ions in neurology and psychiatry* (pp. 165–176). New York: A.R. Liss, Inc.
20. Durlach, J. (1961). Chronic tetany and magnesium depletion. *Lancet*, *1*, 282–283.
21. Durlach, J. (1985). Neurological disturbances due to magnesium imbalance. In S. Gabay, J. Harris, & B.T. Ho (Eds.), *Metal ions in neurology and psychiatry* (pp. 121–128). New York: A.R. Liss, Inc.
22. Erdtman, H. (1927). Glycerophosphatspaltung durch nierenphosphatase und ihore aktivierung. *Z. f. Physiol. Chemie*, *172*, 182–198.
23. Fossel, E.T., Sarasua, M.M., & Koehler, K.A. (1985). A lithium-7 NMR investigation of the lithium ion interaction with phosphatidylcholine-phosphatidylglycerol membranes. *J. Magnetic Resonance*, *64*, 536–540.
24. Frizel, D., Coppen, A., & Marks, V. (1969). Plasma magnesium and calcium in depression. *Br. J. Psychiatry*, *115*, 1375–1377.
25. Glen, A.I.M. (1978). Lithium regulation of membrane ATPases. In F.N. Johnson & S. Johnson (Eds.), *Lithium in medical practice* (pp. 183–192). Lancaster: MTP Press Limited.
26. Gupta, J.D., & Crollini, C. (1975). Effect of lithium on magnesium-dependent enzymes. *Lancet*, *1*, 216–217.
27. Hall, R.C.W., & Jaffee, J.R. (1973). Hypomagnesemia: Physical and psychiatric symptoms. *JAMA*, *224*, 1749–1751.
28. Hallcher, L.M., & Sherman, W.R. (1980). The effects of lithium ions and other agents on the activity of myo-inositol-1-phosphatase from bovine brain. *J. Biol. Chem.*, *255*, 10896–10901.
29. Hertzberg, L., & Hertzberg, B. (1977). Mood change and magnesium. *J. Nerv. Ment. Dis.*, *165*, 423–426.

30. Hokin-Neaverson, M., Buckhardt, W.A., & Jefferson, J.W. (1976). Increased erythrocyte Na⁺ pump and NaK-ATPase activity during lithium therapy. *Res. Commun. Chem. Pathol. Pharmacol.*, *14*, 117–126.

31. Hullin, R.P., Swinscoe, J.C., McDonald, R., & Dransfield, G.A. (1968). Metabolic balance studies on the effect of lithium in manic-depressive psychosis. *Brit. J. Psychiat.*, *114*, 1561–1573.

32. Hullin, R.P. (1975). The effects of lithium on electrolyte balance and body fluids. In F.N. Johnson (Ed.), *Lithium research and therapy* (pp. 359–379). London: Academic Press.

33. Johnson, S. (1975). The effects of lithium on basic cellular processes. In F.N. Johnson (Ed.), *Lithium research and therapy* (pp. 533–556). London: Academic Press.

34. Johnston, B.B., Naylor, G.J., Dick, E.G., Hopwood, S.E., & Dick, D.A.T. (1980). Prediction of clinical course of bipolar manic depressive illness treated with lithium. *Psychol. Med.*, *10*, 329–334.

35. Knapp, S., & Mandell, A.J. (1973). Short and long term lithium administration effects on the brains serotonergic systems. *Science*, *180*, 645–647.

36. Koenig, M.L., & Jope, R.S. (1988). Effects of lithium on synaptosomal Ca^{2+} fluxes. *Psychopharmacology*, *96*, 267–272.

37. Lazarus, J.H. (1986). Lithium and the cell. In J.H. Lazarus (Ed.), *Endocrine and metabolic effects of lithium* (pp. 31–54). New York: Plenum Medical Book Company.

38. Lehman, I.R., Bessman, M.S., Simms, E., & Kronberg, A. (1958). Enzymatic synthesis of deoxyribonucleic acid, I: Preparation of substrates and partial purification of an enzyme from Escherichia coli. *J. Biol. Chem.*, *233*, 160–170.

39. Linnoila, N., MacDonald, E., Reinila, M., LeRoy, A., Rubinow, D.R., & Goodwin, F.K (1983). RBC membrane adenosine trisphosphatase activities in patients with major affective disorders. *Arch. Gen. Psychiatry*, *40*, 1021–1026.

40. MacDonald, E., LeRoy, A., & Linnoila, M. (1982). Failure of lithium to counteract vanadate-induced inhibition of red blood cell membrane Na⁺/K⁺-ATPase. *Lancet*, 774.

41. Maguire, E.M. (1982). Magnesium regulation of the beta-receptor-adenylate cyclase complex. II. Sc^{3+} as an Mg^{2+} antagonist. *Mol. Pharmacol.*, *22*, 274–280.

42. Mandell, A.J. (1978). Redundant mechanisms regulating brain tyrosine and trypthophan hydroxylases. In R. George, R. Okun, & A.K. Cho (Eds.), *Annual review of pharmacology and toxicology* (pp. 461–493). Palo Alto: Annual Reviews Inc.

43. Mandell, A.J., & Knapp, S. (1982). Regulation of trytophan hydroxylase: Variational kinetics suggest a neuropharmacology of phase. In B.T. Ho, J.C. Schoolar, & E. Usdin (Eds.), *Serotonin in biological psychiatry* (pp. 1–15). New York: Raven Press.

44. Mann, A.V. (1977). Resources. In J.O. Bockris (Ed.), *Environmental chemistry* (pp. 137–147). New York: Plenum.

45. Matzen, T.A., & Martin, R.L. (1985). Magnesium deficiency psychosis induced by cancer chemotherapy. *Biol. Psychiatry*, *20*, 788–791.

46. Mellerup, B., & Mellerup E.T. (1984). Seasonal variations in urinary excretion of calcium, magnesium, and phosphate in manic-melancholic patients. *Chronobiol. Int.*, *1*, 81–86.

47. Mellerup, E.T., Lauritsen, B., Dam, H., & Rafaelsen, O.J. (1976). Lithium effects on diurnal rhythms of calcium, magnesium, and phosphate metabolism in manic-melancholic disorder. *Acta Psychiat. Scand.*, *53*, 360–370.

48. Moore, P.B. (1968). Polynucleotide attachment to ribosomes. *J. Mol. Biology*, *18*, 8–20.

49. Mørk, A., & Geisler, A. (1987a). Effects of lithium on calmodulin-stimulated adeny-

late cylase activity in cortical membranes from rat brain. *Pharmacol. Toxicol.*, *60*, 17–23.

50. Mørk, A., & Geisler, A. (1987b). Mode of action of lithium on the catalytic unit of adenylate cyclase from rat brain. *Pharmacol. Toxicol.*, *60*, 241–248.

51. Mørk, A., & Geisler, A. (1989a). Effects of GTP on hormone-stimulated adenylate cyclase activity in cerebral cortex, striatum and hippocampus from rats treated chronically with lithium. *Biol. Psychiatry*, *26*, 279–288.

52. Mørk, A., & Geisler, A. (1989b). The effects of lithium in vitro and ex vivo on adenylate cyclase in brain are exerted by distinct mechanisms. *Neuropharmacology*, *28*, 307–311.

53. Naylor, G.J., Dick, D.A.T., Dick, E.G., LePoidevin, D., & Whyte, S.F. (1972). Erythrocyte membrane cation carrier in depressive illness. *Psychol. Med.*, *3*, 502–508.

54. Naylor, G.J., Dick, D.A.T., Dick, E.G., & Moody, J.P. (1974). Lithium therapy and erythrocyte membrane cation carrier. *Psychopharmacologia*, *37*, 81–86.

55. Naylor, G.J., & Smith, A.H.W. (1981a). Defective genetic control of sodium pump density in manic depressive psychosis. *Psychol. Med.*, *11*, 257–263.

56. Naylor, G.J., & Smith, A.H.W. (1981b). Vanadium, a possible aetiological factor in manic depressive illness. *Psychol. Med.*, *11*, 249–256.

57. Naylor, G.J., Fleming, L.W., Stewart, W.K., McNamee, H.B., & LePoidevin, D. (1972). Plasma magnesium and calcium levels in depressive psychosis. *Br. J. Psychiatry*, *120*, 683–684.

58. Newman, M.E., & Belmaker, R.H. (1987). Effects of lithium in vitro and ex vivo on components of the adenylate cyclase system in membranes from cerebral cortex of the rat. *Neuropharmacology*, *26*, 211–217.

59. Nielsen, J. (1964). Magnesium-lithium studies 1. *Acta Psychiat. Scand.*, *40*,190–196.

60. Ohanian, H., Borhanian, K., de Farias, S., & Bennun, A. (1981). A model for the regulation of brain adenylate cyclase by ionic equilibria. *J. Bioenergics and Biomembranes*, *13*, 317–355.

61. Perez-Cruet, J., Tagliamonte, A., Tagliamonte, P., & Gessa, G.L. (1971). Stimulation of serotonin synthesis by lithium. *J. Pharmacol. Exp. Ther.*, *178*, 325–330.

62. Ragan, C.I., Gee, N., Jackson, R., & Reid, G. (1988). The inhibition by lithium inositol (1,4) bisphosphate metabolism. In N.J. Birch (Ed.), *Lithium: Inorganic pharmacology and psychiatric use* (pp. 205–207). Oxford: IRL Press.

63. Ramsey, T.A., Frazer, A., & Mendels, J. (1979). Plasma and erythrocyte cations in affective illness. *Neuropsychobiology*, *5*, 1–10.

64. Reading, H.W., & Isbir, (1979). Action of lithium on ATPases in the rat iris and visual cortex. *Biochem. Pharmacol.*, *28*, 3471–3474.

65. Schubert, J. (1973). Effects of chronic lithium treatment on monoamine metabolism in rat brain. *Psychopharmacologia*, *32*, 301–311.

66. Schwann, A.C., Henninger, G.R., Roth, R.H., & Maas, J.W. (1981). Differential effects of short and long term lithium on tryptophan uptake on serotonergic cat brain. *Life Sci.*, *28*, 347–354.

67. Seelig, M.S., Berger, A.R., & Spielholz, N. (1975). Latent tetany and anxiety, marginal magnesium deficit, and normal calcemia. *Dis. Nerv. System*, *36*, 461–465.

68. Sengupta, N., Datta, S.C., Sengupta, D., & Bal, S. (1980). Platelet and erythrocyte membrane – adenosine triphosphatase activity in depressive and manic-depressive illness. *Psychiatry Res.*, *3*, 337–344.

69. Sulakhe, P.V., & Höehn, E.K. (1984). Interaction of EGTA with a hydrophobic region inhibits particulate adenylate cyclase from rat cerebral cortex: A study of an EGTA-inhibitable enzyme by using alamethicin. *Int. J. Biochem.*, *16*, 1029–1035.
70. Sulser, F., Janowsky, A.J., Okada, F., Manier, D.H., & Mobly, P.L. (1983). Regulation of recognition and action function of the norepinephrine (NE) receptor-coupled adenylate cyclase system in brain: Implications for the therapy of depression. *Neuropharmacology*, *22*, 425–431.
71. Swanson, P.D., Anderson, L., & Stahl, W.L. (1974). Uptake of Ca ions by synaptosomes from rat brain. *Biochim. Biophys. Acta*, *356*, 174–183.
72. Wacker, W.E.C., & Parisi, A.F. (1968). Magnesium metabolism. *N. Eng. J. Med.*, *278*, 712–717.
73. Wacker, W.E.C. (1980). Biochemistry and physiology. In W.E.C. Wacker (Ed.), *Magnesium and man* (pp. 11–51). Massachusetts: Harvard University Press.

11
Effects of Lithium on Essential Fatty Acid and Prostaglandin Metabolism

DAVID F. HORROBIN

Introduction

Lithium has an established therapeutic effect in the treatment and prevention of manic-depressive psychosis. This effect is exerted at plasma lithium concentrations over the range of about 0.5 to 1.5 mM. In recent years, partly due to concerns about the toxic effects of higher levels, and partly due to recognition that the higher concentrations may not be required to achieve a therapeutic response, there has been a trend toward trying to maintain plasma concentrations at the lower end of this range.

As yet, there is no agreement concerning the biochemical effect of lithium which is responsible for the therapeutic effect. It is possible, even likely, that no single mechanism is involved but that several different actions contribute to the final result. However, in order to be seriously considered as a candidate, the effect must be demonstrable at concentrations not greatly in excess of the normal therapeutic range. Moreover, there should be a plausible relationship between the effect observed and the psychiatric phenomena which are being influenced.

Recently, a quite different category of therapeutic actions of lithium has been demonstrated. Topical applications of lithium ions have been found to be clinically effective in treating a variety of inflammatory skin disorders, including various forms of dermatitis and pruritus (1), genital herpes (2), and seborrhoeic dermatitis (3). Because the lithium ions are administered in various types of bases, and because the rate of penetration into the skin is unknown, it is difficult at present to know precisely the therapeutic concentration at the site of action. However, it seems probable that it is in the region of 50 to 200 mM.

Thus, in explaining the therapeutic effects of lithium ions, we are looking for effects at two quite different concentration ranges, below 1 mM and above 50 mM. Most early theories of lithium action in manic depression concentrated on its ability to interfere with movements of sodium and potassium ions (4). Recently there has been less interest in this approach, as it has become apparent that the lithium concentrations which are required to exert any appreciable effect are substantially higher than those present in the blood of lithium-treated manic-depressive patients. However, the concentrations at which lithium displaces sodium and potassium are relevant to the topical anti-inflammatory effects of lithium. This approach to the mechanism of lithium action should be re-examined in relation to anti-inflammatory effects.

There are at present four main candidates for the mechanisms involved in those effects of lithium that are demonstrable at concentrations of 1 mM or less. It is possible, indeed probable, that when these mechanisms are finally understood they will be seen not to be independent but to be mutually interdependent aspects of the same fundamental process. These effects are:

1. Lithium regulation of adenylate cyclase activity especially as stimulated by agonists such as prostaglandin (PG) E_1 (5–8). Many hormones and neurotransmitters exert their effects at least in part by stimulating the synthesis of cyclic AMP. Lithium can interfere with this stimulation and at least some of the experiments have shown that it can do so at concentrations that are relevant to its psychiatric effects.
2. Lithium regulation of inositol phospholipid metabolism, whereby lithium elevates myo-inositol-1-phosphate and lowers free inositol concentrations (9–13). Modulation of inositol phospholipid metabolism is increasingly being recognized as a central process in many mechanisms of cell activation.
3. Lithium regulation of G-protein receptor coupling, whereby lithium can reduce the coupling of activated receptors to the relevant G-proteins (14). G-proteins are important links between receptor activation and second messenger systems, including those which involve regulation of inositol metabolism.
4. Lithium regulation of the production of eicosanoids from their essential fatty acid precursors. This is the subject of this review.

It is likely that these four mechanisms are interrelated. For example, the regulation of the synthesis of prostaglandins of the E series by lithium would modulate cyclic AMP concentrations, since E prostaglandins are important determinants of adenyl cyclase activity. Inositol phospholipids are important reservoirs of the essential fatty acids from which the prostaglandins and other eicosanoids are derived. myo-Inositol itself is a potent stimulator of PG formation from arachidonic acid (15), and there are important interactions between the arachidonic acid and inositol regulating systems (16,17). G-proteins are important links between receptor activation and second messenger systems of many types.

The mechanisms of lithium action other than on the essential fatty acid (EFA)/eicosanoid system are reviewed in detail in other chapters of this book. This chapter concentrates on the EFAs and the eicosanoids. However, in reading this and the other chapters it is important to remember that there are likely to be currently undemonstrated links between these various mechanisms of action.

The Essential Fatty Acids and Eicosanoids

Linoleic acid is an essential nutrient, an essential fatty acid (EFA). But in order to be fully utilized by the body, linoleic acid must be metabolized to a variety of other EFAs and to their derivatives, the eicosanoids. The eicosanoids include the prostaglandins (PGs) and a variety of other biologically active molecules, including the leukotrienes and various hydroxy-fatty acids, which may be derived from the EFAs by cyclo-oxygenase, lipoxygenase, and other enzyme systems. The metabolic pathways involved are exceedingly complex and scores of metabolic products of linoleic acid have now been identified. A detailed description of the biochemistry involved is beyond the scope of this chapter but the pathway in outline is shown in Figure 11.1.

The EFAs and eicosanoids have two broad groups of functions. The first is structural. The EFAs are components of every membrane in the body and because of their unsaturation may have profound effects on fluidity, flexibility, and precise membrane structure (18). The EFAs are important determinants of the properties of the lipid environment in which membrane-associated proteins such as receptors, enzymes, and ion channels lie embedded. The EFAs therefore are able to exert subtle modulating actions on the behavior of these proteins. The second function of the EFA metabolites is regulatory. There is a bewildering array of eicosanoids that can be very rapidly produced and almost equally rapidly destroyed in response to the immediate needs of the cell concerned. These eicosanoids can influence the activity of almost every system in which they have been tested. Effects include modulation of smooth muscle, of the behavior of cells involved in inflammatory and immune responses, of nerve conduction, of transmitter release and action, and of platelet reactivity. Each eicosanoid has a specific effect on each tissue. Extrapolation from one tissue to another is frequently unjustified and should be done only with full awareness of the problems involved.

One of the eicosanoid effects that has caused great confusion but which may be highly relevant to the mechanism of action of lithium is the phenomenon of bell-shaped dose-response curves (19,20). Often an eicosanoid will have an effect which, as the eicosanoid concentration rises, also rises to a peak. There is nothing surprising about that. But what is surprising is that, as the concentration of the eicosanoid rises further, the maximum effect is seen to be a peak and not a plateau: the effect disappears with increasing concentration. This type of action is particularly apparent with PGE_1 in a variety of circumstances (19), including its effects on smooth muscle (21) and on the immune system (22,23).

The regulation of eicosanoid formation is complex and depends on a number of factors. The two main precursors are dihomogammalinolenic acid (DGLA) which gives rise to 1 series PGs, and arachidonic acid (AA), which gives rise to 2 series PGs. The cyclo-oxygenase and lipoxygenase enzyme systems which metabolize these precursors can probably act on the EFAs only when the latter are in their free form. Free EFAs are found only in very small amounts under physiological conditions, the bulk being covalently linked to phospholipids, cholesterol esters, and triglycerides. Release of the free EFA is therefore the first step in eicosanoid production. While it is widely assumed that the phospholipids are the only sources of EFAs for eicosanoid formation, recent observations suggest that under appropriate conditions both triglycerides and cholesterol esters, and even fatty acids formed rapidly from dietary EFAs, can act as precursor pools (24–26). EFAs are released from phospholipids by either phospholipase A_2 or phospholipase C, and from triglycerides by lipases and cholesterol esters by esterases.

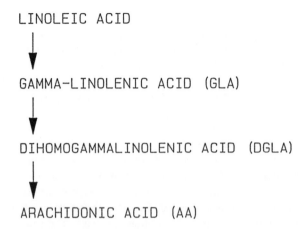

FIGURE 11.1a. Outline of the pathway of linoleic acid metabolism.

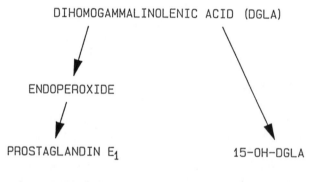

FIGURE 11.1b. The metabolism of dihomogammalinolenic acid to prostaglandin E_1 and 15-hydroxy-dihomogammalinolenic acid.

What determines whether a particular free EFA molecule will be metabolized by the cyclo-oxygenase route to prostaglandins, prostacyclin, and thromboxanes, or by the lipoxygenase route to hydroxy-acids and leukotrienes is as yet unclear. There is some evidence that selective inhibition of one enzyme will lead to increased formation of the products of the other enzyme but the relationship is by no means 1 to 1. The levels of the particular enzyme within the cell and other unknown regulating factors must play roles. Equally, the factors which determine the nature of the final products of the two enzyme systems are uncertain. The unstable endoperoxide intermediates of the cyclo-oxygenase system may be converted to many different end products. This partly depends on the presence within that particular cell of isomerizing enzymes which, for example, direct the synthesis from arachidonic acid of PGE_2 or PGD_2 or thromboxane A_2 or

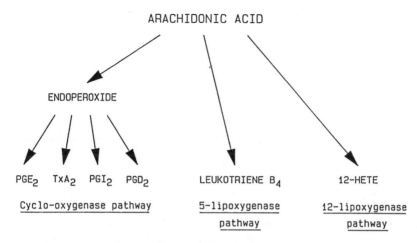

FIGURE 11.1c. The main metabolites of arachidonic acid. TxA$_2$ = thromboxane A$_2$; PGI$_2$ = prostacyclin.

prostacyclin. But how these isomerases are regulated is at present something of a mystery.

It is important to emphasize that the overall results of metabolism of the two main eicosanoid precursors, DGLA and AA, are quite different. Major metabolites of DGLA appear to be PGE$_1$ and 15-OH-DGLA (27,28). PGE$_1$ has a wide variety of desirable effects (29,30), while 15-OH-DGLA inhibits the metabolism of AA by both 5- and 12-lipoxygenase systems. This limits the production of potentially harmful leukotrienes and hydroxy-acids which may be involved in many types of inflammation and smooth muscle contraction. The effects of PGE$_1$ include inhibition of phospholipase A$_2$, inhibition of platelet aggregation, relaxation of smooth muscle, lowering of blood pressure, and inhibition of cholesterol biosynthesis. There appears to be a reciprocal relationship between the products of DGLA and of AA metabolism (22,29,31,32). Elevation of formation of DGLA products will tend to inhibit the formation of AA products, while a lack of DGLA products will lead to increased conversion of AA to its potentially harmful metabolites.

Eicosanoids in Affective Disorders

There are good reasons for considering the possibility that abnormalities of EFA and eicosanoid metabolism may be important in affective disorders and related conditions such as alcoholism (33–35). First, the dry weight of the brain is about 20% EFA. Second, the EFAs in the cell membrane will influence nerve conduction, neurotransmitter release, and neurotransmitter action. Third, the eicosanoids produced from the EFAs may have profound effects on many aspects of neuronal function (33–35).

Unfortunately, as yet the actual evidence relating EFAs to affective disorders is sparse. Measurements of eicosanoids are difficult, especially at the very low concentrations which are biologically active. Moreover, it is usually difficult to measure changes in more than one or two eicosanoids at the same time: important changes in the levels of many other eicosanoids may therefore be missed.

In depressed individuals, conversion of DGLA to PGE_1 by platelets is reduced, whereas in manic individuals it is elevated (36). Also in depression, plasma levels of the AA metabolites, PGE_2 and thromboxane B_2 (the stable metabolite of thromboxane A_2) and cerebrospinal fluid levels of a metabolite of PGE_2 have been found to be elevated (37,38). Different forms of stress in humans have been shown to elevate blood concentrations of PGE_2 and thromboxane B_2 (39).

The reduced formation of PGE_1 by platelets, and the elevated levels of 2 series eicosanoids in body fluids in depression are interesting because of the reciprocal relationship which exists between DGLA and arachidonic acid. A reduction of the formation of DGLA metabolites is likely to lead to an increase in production of the metabolites derived from AA.

Other observations are consistent with the idea that, directly or indirectly, PGE_1 may be an important regulator of mood. PGE_1 infusions are now used in some centers in the treatment of peripheral vascular disease. Although not formally described in the literature, several physicians administering such infusions have reported to me that their patients appear to experience an elevation of mood during and after the infusion. This elevation of mood does not appear to be secondary to the peripheral vasodilatation. Prostacyclin, the AA metabolite, which produces an equivalent peripheral vasodilatation is not associated with similar mood elevation: if anything, prostacyclin tends to cause dysphoria. Drinking alcohol is also commonly found to be an elevator of mood. Levels of alcohol associated with mild intoxication have been shown to stimulate the formation of PGE_1 by normal human platelets (33,34,40).

It has therefore been proposed that in mania there may be a general elevation of the production of PGE_1 with a consequent suppression of 2 series eicosanoids, whereas in depression the reverse situation may exist (33,34). Since PGE_1 has anti-inflammatory and mildly immunosuppressing actions, whereas the 2 series eicosanoids are consistently pro-inflammatory (22,31,32), this may help to explain why depression appears to be associated with a higher incidence than normal of inflammatory and auto-immune disorders, while mania may be associated with suppression of those disorders (41,42).

Lithium Effects on Eicosanoids

Unfortunately, our knowledge of lithium effects on eicosanoid metabolism is confined to its short-term actions in one system, the perfused superior mesenteric artery bed of the rat. However, the experimental demonstrations of lithium actions of other types are also similarly limited in the range of systems which have been investigated. It is hoped that others may extend the observations we have made to other types of systems.

The first observations of lithium actions on EFAs and eicosanoids were indirect. In the rat mesentery, prolactin and zinc seem to exert their actions in part by mobilizing DGLA and increasing 1 series PG formation: vasopressin in contrast acts by mobilizing AA an stimulating 2 series eicosanoid formation (43,44). Lithium, at clinically relevant concentrations, inhibited the actions of prolactin and zinc, but not that of vasopressin. At higher concentrations lithium blocked the effects of both prolactin and vasopressin. This suggests that concentrations of lithium relevant to psychiatry may selectively inhibit the formation of 1 series eicosanoids. Higher concentrations may inhibit the formation of both types of prostaglandins.

More recently, the outputs of DGLA and its metabolite PGE_1, and of AA and its metabolites PGE_2, thromboxane B_2 and prostacyclin (measured as its stable metabolite, 6-keto-PGF_1) have been directly measured in the effluent from the perfused mesenteric vascular bed (45). A range of lithium concentrations has been tested. The results obtained at 0.8 mM lithium, clearly relevant to the action of lithium in manic-depression, are shown in Figure 11.2. Results at 0.4 and 1.6 mM were quantitatively different but qualitatively similar. They may be summarized as follows:

1. Lithium has a modest, but statistically significant, effect on inhibiting release of DGLA from the tissue. It has a much greater effect on inhibiting PGE_1 formation. This suggests that there may be two distinct actions of lithium, one on the mobilization of DGLA from its various stores, and one on the conversion of the free DGLA to PGE_1.
2. Lithium has nonsignificant actions on the release of AA, prostacyclin, and PGE_2, but a strong inhibitory effect on thromboxane B_2, the stable metabolite of thromboxane A_2. This effect on TxA_2 is probably related to an action on the conversion of the unstable endoperoxide to the thromboxane. The absence of any effects on AA, PGE_2 or prostacyclin, makes it unlikely that the lithium action is exerted earlier in the pathway.

It must again be emphasized that these actions of lithium have been demonstrated in short-term in vitro experiments in one system only. It is uncertain whether they can be applied to other tissues, or what the effects of lithium might be in the longer term. However, it is interesting that these effects offer a plausible explanation of the antimanic and antidepressive effects of lithium.

EFAs, Prostaglandins, and Manic-Depression

There is evidence, briefly reviewed earlier, that manic and euphoric states may be associated with excess formation of PGE_1. On the other hand, depression may be associated with reduced formation of PGE_1 and excess formation of both thromboxane A_2 and PGE_2. Lithium, at clinically relevant concentrations, has an inhibitory effect on the formation of both PGE_1 and thromboxane A_2. The effect on the formation of PGE_1 could be related to the antimanic action of lithium, while the effect on TxA_2 could be associated with the antidepressive action of

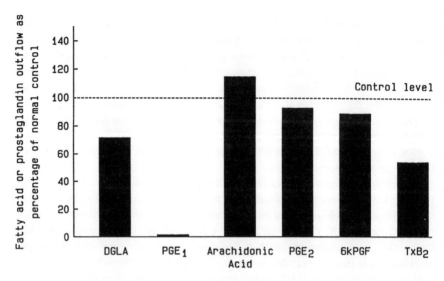

FIGURE 11.2. The effects of 0.8 mM lithium chloride on the release of essential fatty acids and prostaglandins into the effluent from the buffer-perfused superior mesenteric artery vascular bed in the rat. The bars indicate the levels when lithium was added to the buffer as compared to the levels with no lithium (indicated by the *dotted line*). There were eight animals in the control and eight in the lithium-treated group. Lithium significantly reduced the outflow of dihomogammalinolenic acid (DGLA), PGE_1, and thromboxane B_2 (the stable metabolite of TxA_2). It had no significant effect on the outflow of arachidonic acid, PGE_2 or 6-keto-PGF_1, the stable metabolite of PGI_2 (prostacyclin).

lithium. Lithium has no effects on PGE_2, also elevated in depression, which might help to account for its relative inefficacy in treating existing depression as opposed to its effectiveness in treating mania.

In contrast to its weak actions in treating depression, lithium is very effective in preventing depressive episodes in patients with bipolar affective disorders. This dual effect is not satisfactorily explained by any of the current theories of lithium action. However, the observed effects of lithium on EFA and prostaglandin biochemistry offer a possible explanation. The relatively scanty data suggest that mania is associated with excessive production of 1 series PGs, while depression is associated with inadequate formation of 1 series PGs and an excess of 2 series PGs. As mentioned earlier, there appears to be a reciprocal relationship between the formation of the 1 and 2 series products: high levels of PGE_1 inhibit the formation of 2 series PGs, while low levels of PGE_1 are associated with excessive production of 2 series PGs.

Suppose that mania is associated with unregulated excessive conversion of DGLA to PGE_1. Stores of DGLA in the body are limited, and if prolonged such an unregulated flow of PGE_1 could lead to depletion of its precursor, and hence a collapse of PGE_1 biosynthesis. The effect might be likened to opening the outflow valve on a water reservoir with limited inflow: initially the flow of water would be greatly increased, but would fall drastically again once the reservoir became empty. Thus a period of excessive PGE_1 formation would be followed by a drastic fall: this in itself might lead to depressed mood, but it would also lead to increased production of 2 series PGs because of the previously mentioned reciprocal relationship. Such increased production is known to be associated with depression.

Given this mechanism, it is possible to see how a modest inhibition of PGE_1 formation by lithium could prevent both mania and depression. A level of lithium which slowed the conversion of DGLA to PGE_1, without stopping it completely, would prevent excess PGE_1 formation. If that excess PGE_1 formation were contributing to the mania, the mania would be treated or prevented. The same level of lithium would also prevent depletion of DGLA, allowing the DGLA stores to be replenished at a rate comparable to that at which they were being used up. Thus lithium would prevent the catastrophic fall in PGE_1 production which might occur consequent upon depletion of DGLA stores. As a result it would also prevent the excessive conversion of AA to its eicosanoid metabolites and the development of depression.

The prostaglandin hypothesis is therefore unusual among the biochemical mechanisms proposed to explain lithium action in providing an account of the way in which lithium may be effective in preventing both elevated and depressed mood states.

Anti-Inflammatory Effects of Lithium

There is abundant evidence from a great number of studies showing that overproduction of eicosanoids derived from arachidonic acid is involved in many types of inflammation. Steroids are believed to exert their anti-inflammatory effects by inhibiting mobilization of arachidonic acid by phospholipase A_2: in consequence they reduce production of both lipoxygenase and cyclo-oxygenase metabolites. Nonsteroidal anti-inflammatory drugs (NSAIDS) are believed to work by inhibiting the cyclo-oxygenase system: they usually do not block the lipoxygenases which is believed to explain why they are less potent than the steroids as anti-inflammatory agents.

As mentioned in the introduction, there is increasing evidence that lithium exerts anti-inflammatory effects on a variety of skin conditions. These effects probably require local lithium concentrations in excess of 50 mM. Using the same perfused mesenteric vascular bed model, the effects of the high concentrations of lithium which are relevant to topical application have been investigated (45). The high concentrations of lithium were found to inhibit the release from the tissue of both DGLA and AA. They also produced dramatic and highly significant inhibition of the formation of all the four eicosanoids measured; of PGE_1 by 85%; of PGE_2 by 72%; of TxA_2 by 77%; and of prostacyclin by 69%. These results indicate that lithium probably blocks the synthesis of all the cyclo-oxygenase metabolites of all the EFA precursors of PGs. As yet nothing is known of the effects on lipoxygenase metabolites. However, the inhibitory effects of lithium at high concentrations on DGLA and AA release suggest that production of lipoxygenase metabolites is also likely to be reduced.

These effects of lithium on prostaglandin and EFA metabolism are therefore consistent with the idea that the anti-inflammatory effects of high concentrations of lithium can be accounted for on this basis.

Lithium Actions in Other Systems

There are several other systems in which the effects of lithium on EFA and prostaglandin metabolism may be relevant to the clinical results observed. Lithium is known to stimulate neutrophil formation in patients treated with chemotherapy for malignant disease, and also in grey collie dogs suffering from cyclic haematopoiesis (46,47). Prostaglandins are known to be important in regulating leukocyte formation and may be involved in controlling the effects of colony-stimulating factors. The relationships between Lithium and prostaglandin effects on hematopoiesis are discussed in detail (see Chapter 6 by Gallicchio). Lithium regulation of eicosanoid biochemistry could therefore be important in these actions on neutrophil formation and action.

Lithium has also been shown to have many effects on the immune system (41,42,47). Interestingly, these have a biphasic nature, similar to the actions of lithium on mania and depression. In some circumstances lithium may be able to activate immune function, while in others it seems to damp down excessive immunological reactions. Since PGE_1 has similar biphasic actions on immunity, the interaction between lithium and immune function is worth further exploration in relation to the EFA and eicosanoid system. However, the effects of both lithium and the EFAs and eicosanoids on immune function are so multifarious and complex that it is at present difficult to construct plausible models. A good deal more information is required. (See chapter by Hart)

Conclusions

The known effects of lithium on EFA and prostaglandin metabolism offer plausible explanations for its actions at concentrations up to 1 mM on manic depression, and for its actions at concentrations two orders of magnitude higher on cutaneous inflammation. It is likely that these effects on eicosanoids will be found to be closely related to actions on adenylate cyclase, on phosphatidyl-1-inositol metabolism and on G-proteins. With all these proposed mechanisms of action, one unfortunate feature is that experimental results have been obtained in very limited numbers of systems. These experimental systems are of uncertain relevance to the known targets of lithium action in clinical medicine. It is important that serious attempts be made to relate the experimental data to the clinical situation.

References

1. Sherwin, L. (1974). Compositions containing lithium succinate. *US Patent, 3*, 639, 625.
2. Skinner, G.R.B. (1983). Lithium ointment for genital herpes. *Lancet, 2*, 288.
3. Boyle, J., Burton, J.L., & Faergemann, J. (1986). Use of topical lithium succinate for seborrhoeic dermatitis. *Br. Med. J., 292*, 28.
4. Maletzky, B., & Blachly, P.H. (1971). *The use of lithium in psychiatry*. London: Butterworths.
5. Pandey, G.N., & Davis, J.M. (1979). Cyclic AMP and adenylate cyclase in psychiatric illness. In G.C. Palmer (Ed.), Neuropharmacology of cyclic nucleotides. Baltimore: Urban & Schwarzenberg.
6. Wolff, J., Berens, S.C., & Jones, A.B. (1970). Inhibition of thyrotrophin-stimulated adenyl cyclase activity of beef thyroid membranes by low concentrations of lithium ion. *Biochem. Biophys. Res. Comm., 39*, 77–82.
7. Dousa, T.P. (1974). Interaction of lithium with vasopressin-sensitive cyclic AMP system of human renal medulla. *Endocrinology, 95*, 1359–1366.
8. Ebstein, R.P., Belmaker, R.H., Brunhaus, L., et al. (1976). Lithium inhibition of adrenaline sensitive adenylate cyclase in humans. *Nature, 259*, 411–413.
9. Hallcher, L.M., & Sherman, W.R. (1980). The effects of lithium ion and other agents on the activity of myo-inositol-1-phosphatase from bovine brain. *J. Biol. Chem., 255*, 10896–10901.
10. Berridge, M.J., Downes, C.P., & Hanley, M.R. (1982). Lithium amplifies agonist-dependent phosphatidylinositol responses in brain and salivary glands. *Biochem. J., 206*, 587–595.
11. Worley, P.F., Heller, W.A., Snyder, S.H., et al. (1988). Lithium blocks a phosphoinositide-mediated cholinergic response in hippocampal slices. *Science, 239*, 1428–1429.
12. Shears, S.B. (1988). Lithium and inositol lipid turnover. In N.J. Birch (Ed.), *Lithium: Inorganic pharmacology and psychiatric use*. Oxford: IRL Press.
13. Ragan, C.I., Gee, N., Jackson, R., et al. (1988). The inhibition by lithium of inositol (1,4) bisphosphate metabolism. In N.J. Birch (Ed.), *Lithium: Inorganic pharmacology and psychiatric use*. Oxford: IRL Press.
14. Avissar, S., Schreiber, G., Danon, A., et al. (1988). Lithium inhibits adrenergic and cholinergic increases in GTP binding in rat cortex. *Nature, 331*, 440–442.
15. Karmazyn, M., Horrobin, D.F., Manku, M.S., et al. (1977). myo-Inositol in physiological concentrations stimulates production of PG-like material. *Prostaglandins, 14*, 967–974.
16. Flint, A.P., Leat, W.M., Sheldrick, E.L., et al. (1986). Stimulation of phosphoinositide hydrolysis by oxytocin and the mechanism by which oxytocin controls prostaglandin synthesis in the ovine endometrium. *Biochem. J., 237*, 797–805.
17. Chaudhry, A., Laychock, S.G., & Rubin, R.P. (1987). The effects of fatty acids on phosphoinositide synthesis and myo-inositol accumulation in exocrine pancreas. *J. Biol. Chem., 262*, 17426–17431.
18. McMurchie, E.J. (1988). Dietary lipids and the regulation of membrane fluidity and function. In *Physiological regulation of membrane fluidity*. New York: Alan R. Liss.
19. Horrobin, D.F. (1978). *Prostaglandins: Physiology, pharmacology and clinical significance*. Edinburgh: Churchill Livingstone.

20. Horrobin, D.F. (1977). Interactions between prostaglandins and calcium: The importance of bell-shaped dose-response curves. *Prostaglandins, 14*, 667–677.
21. Manku, M.S., Mtabaji, J.P., & Horrobin, D.F. (1977). Effects of prostaglandins on baseline pressure and responses to noradrenaline in a perfused rat mesenteric artery preparation: PGE_1 as an antagonist of PGE_2. *Prostaglandins, 13*, 701–709.
22. Horrobin, D.F. (1980). The regulation of prostaglandin biosynthesis: Negative feedback mechanisms and the selective control of formation of 1 and 2 series prostaglandins: Relevance to inflammation and immunity. *Med. Hypotheses, 6*, 687–709.
23. Mertin, J., & Stackpoole, A. (1981). Anti-PGE antibodies inhibit in vivo development of cell-mediated immunity. *Nature, 294*, 456–458.
24. Vahouny, G.V., Chanderbhan, R., Bisgaier, C., et al. (1981). Essential fatty acids and adrenal steroidogenesis. *Progr. Lipid Res., 20*, 233–240.
25. Fujimoto, Y., Nishioka, K., Hase, Y., et al. (1988). Triacylglycerol lipase-mediated release of arachidonic acid for renal medullary prostaglandin synthesis. *Arch. Biochem. Biophys., 261*, 368–374.
26. Mtabaji, J.P., Manku, M.S., & Horrobin, D.F. (1988). Release of fatty acids by perfused vascular tissue in normotensive and hypertensive rats. *Hypertension, 12*, 39–45.
27. Miller, C.C., McCready, C.A., Jones, A.D., & Ziboh, V.A. (1988). Oxidative metabolism of dihomogammalinolenic acid by guinea pig epidermis: Evidence of generation of anti-inflammatory products. *Prostaglandins, 35*, 917–937.
28. Miller, C.C., & Ziboh, V.A. (1988). Gamma-linolenic acid enriched diet alters cutaneous eicosanoids. *Biochem. Biophys. Res. Comm., 154*, 967–974.
29. Horrobin, D.F. (1988). Prostaglandin E_1: Physiological significance and clinical use. *Wien Klin Wochenschr, 100*, 472–477.
30. Kirtland, S.J. (1988). Prostaglandin E_1: A review. *Prostaglandins Leukotr Ess Fatty Acids, 32*, 165–174.
31. Horrobin, D.F., Manku, M.S., & Huang, Y-S. (1984). Effects of essential fatty acids on prostaglandin biosynthesis. *Biomed. Biochim. Acta, 43*, S114–S120.
32. Horrobin, D.F. (1984). Essential fatty acid metabolism in diseases of connective tissue with special reference to scleroderma and to Sjogren's syndrome. *Med. Hypotheses, 14*, 233–247.
33. Horrobin, D.F., & Manku, M.S. (1980). Possible role of prostaglandin E_1 in the affective disorders and in alcoholism. *Br. Med. J., 280*, 1363–1366.
34. Horrobin, D.F. (1987). Essential fatty acids, prostaglandins and alcoholism: An overview. *Alcoholism: Clin. Exp. Res. 11*, 2–9.
35. Horrobin, D.F. (1985). Essential fatty acids and prostaglandins in schizophrenia and alcoholism. In C. Shagass (Ed.), *Biological psychiatry.* Amsterdam: Elsevier.
36. Abdulla, Y.H., & Hamadah, K. (1975). Effect of ADP on PGE formation in blood platelets from patients with depression, mania and schizophrenia. *Br. J. Psychiatry, 127*, 591–595.
37. Lieb, J., Karmali, R.A., & Horrobin, D.F. (1983). Elevated levels of prostaglandin E_2 and thromboxane B_2 in depression. *Prostaglandins Leukotr. Med., 10*, 361–367.
38. Calabrese, J.R., Skwerer, R.G., Barna, B., et al. (1986). Depression, immunocompetence and prostaglandins of the E series. *Psychiatry Res., 17*, 41–47.
39. Mest, H-J., Zehl, U., Sziegoleit, W., et al. (1982). Influence of mental stress on plasma level of prostaglandins, thromboxane B2 and on circulating platelet aggregates in man. *Prostaglandins Leukotr. Med., 8*, 553–563.
40. Manku, M.S., Oka, M., & Horrobin, D.F. (1979). Differential regulation of the formation of prostaglandins and related substances from arachidonic acid and

from dihomogammalinolenic acid. I. Effects of ethanol. *Prostaglandins Med.*, *3*, 119–128.

41. Horrobin, D.F., & Lieb, J. (1981). A biochemical basis for the actions of lithium on behaviour and on immunity: Relapsing and remitting disorders of inflammation and immunity such as multiple sclerosis or recurrent herpes as manic-depression of the immune system. *Med. Hypotheses*, *7*, 891–905.
42. Lieb, J. (1981). Immunopotentiation and inhibition of herpes virus activation during therapy with lithium carbonate. *Med. Hypotheses*, *7*, 885–890.
43. Manku, M.S., Horrobin, D.F., Karmazyn, M., et al. (1979). Prolactin and zinc effects on rat vascular reactivity: Possible relationship to dihomogammalinolenic acid and to prostaglandin synthesis. *Endocrinology*, *104*, 774–779.
44. Karmazyn, M., Manku, M.S., & Horrobin, D.F. (1978). Changes of vascular reactivity induced by low vasopressin concentrations: Interactions with cortisol and lithium and possible involvement of prostaglandins. *Endocrinology*, *102*, 1230–1236.
45. Horrobin, D.F., Jenkins, D.K., Mitchell, J., & Manku, M.S. (1988). Lithium effects on essential fatty acid and prostaglandin metabolism. In N.J. Birch (Ed.), Lithium: Inorganic pharmacology and psychiatric use. Oxford: IRL Press.
46. Hammond, W.P., & Dale, D.C. (1980). Lithium therapy of canine cyclic hematopoiesis. *Blood*, *55*, 26–28.
47. Hart, D.A. (1986). Lithium as an in vitro modulator of immune cell function: Clues to its in vivo biological activities. *IRCS Med. Sci.*, *14*, 756–762.

12
Lithium and Virus Infections

Steven Specter, Gerald Lancz, and Ricardo O. Bach

Introduction

Lithium salts (Li) have been shown to influence viral infections by two distinct mechanisms: by inhibiting the cycle of virus replication and by stimulating protective host immune responses to viruses. Refer to Chapter 5 for an in depth discussion of stimulation of immune cell functions. Lithium has been shown to have anti-inflammatory effects (1), to enhance both lymphocyte and macrophage functions (2), to decrease the activity of suppressor T lymphocytes (3), and to enhance granulocyte function (4). The in vitro and in vivo antiviral activities of lithium are the focus of this chapter.

The earliest indication that lithium has a palliative effect on viral infections comes from a report by Lieb that manic depressive patients undergoing lithium treatment had fewer recurrences of herpes simplex virus (HSV) infection than did patients not receiving lithium (5). Subsequent to this report Skinner and co-workers reported that individuals who used topically applied ointment that contained 8% lithium succinate, 0.05% zinc sulfate, and 0.1% D, L-α-tocopherol reported reduced incidence and severity of recurrent genital HSV infection (6). [Topical application was begun within 2 days of the onset of lesions and ointment was applied 4 times per day for 7 days.] Horrobin reports that he, Skinner, Bach, and Lieb all have observed palliative effects of lithium for HSV and recurrent varicella-zoster virus infections (shingles) (7); however, these observations, other than the report by Skinner (6), are anecdotal. More recently, Parks et al. reported a double blind study using but a single patient who received an alternating course of therapy for 3 months with lithium carbonate (600 mg per day) or placebo for recurrent genital HSV 2 infection (8). They reported that while on placebo the patient experienced 7 recurrences which lasted for a total of 32 days, and during the period of lithium carbonate therapy there was only 1 recurrence lasting for 4 days. Although only one patient was used, they report a statistically significant reduction in genital HSV2 recurrences. While this study was very limited in scope, it provides evidence that lithium salts may serve as antiherpes agents.

The acquired immunodeficiency syndrome (AIDS) epidemic has stirred interest in the antiviral activity of lithium as a potentially useful drug with biological activity against human immunodeficiency virus (HIV), the infectious agent which has the central etiologic role in this syndrome (9,10). Parenti et al. did not observe any significant palliative effects associated with the treatment of 10 HIV-infected homosexual men (9 of them without AIDS-related symptoms) with lithium carbonate per os (9), although 6 subjects in the study stopped treatment (dropped out of the study) due to the development of lithium toxicity. One subject did show a 4-fold decrease in viral titer measured in venous blood 8 weeks after therapy began. The significance of this observation, if any, remains unknown. The 4 subjects that remained in the study did not exhibit any beneficial or detrimental changes in the immune parameters examined during the course of the study. The duration of this study was too short and the doses of lithium carbonate may have been too high (based on the high number of subjects who exhibited toxic levels of lithium carbonate in their blood) to derive significant conclusions from this study. Thus, the authors conclusion that, "there appears no rationale for the use of lithium carbonate in patients with HIV-related dysfunction," (9) may be premature. Herbert, Jacobson, et al. report treating AIDS patients with lithium salt in conjunction with azidothymidine (AZT) and compared this treatment regimen to administering AZT alone (10), the latter being the most frequently employed antiviral therapy currently available for AIDS. Results of this study have as yet not been reported. The potential for using lithium in conjunction with AZT may be related to its ability to stimulate granulocyte production (9). This is significant since granulocytopenia caused by AZT has resulted in cessation of its use by AIDS patients. This and the other immunostimulatory properties mentioned earlier and discussed in Chapter 5 suggest potentially beneficial effects of lithium in AIDS therapy.

In Vitro Effect of Lithium on Viruses

Skinner first reported that lithium would inhibit replication of HSV in mammalian cells cultured in vitro (11,12). Concentrations of 5 to 30 mM lithium chloride (LiCl) inhibited viral replication. HSV, vaccinia virus, and pseudorabies virus, all of which are DNA viruses, were inhibited when lithium was present in virus-infected cell cultures. Two RNA viruses, influenza virus and encephalomyocarditis virus, were not inhibited by lithium. These few studies imply that the action of lithium as an antiviral agent may be limited to DNA viruses, or perhaps to viruses that utilize a DNA intermediate in their replicative cycles. Skinner noted that the inhibition of virus production seemed related to an inhibition of viral DNA polymerase production or activity. Unfortunately, these observations apparently have not been verified or extended by other investigators or additional experimentation.

Studies initiated in our laboratory examined the effects of LiCl, at concentrations of 10 to 40 mM, on the replication of two members of the herpes group of viruses, HSV and varicella-zoster virus. Our results confirm and extend Skinner's observations (13,14). Our studies were performed using two human cell lines, MRC5 and foreskin fibroblasts, as well as rabbit skin fibroblasts (HSV

TABLE 12.1. Inhibition of the development of herpes simplex virus associated cytopathology in vitro by lithium chloride.

Virus added (PFU)[c]	Day post infection	Viral cytopathogenic effects[a] Lithium chloride concentration (mM)			
		0	10	20	40[b]
10^2	1	+	±	0	±
	3	+++	++	+	+
	7	++++	+++	++	++
10^3	1	+	±	±	±
	3	+++	++	++	+
	7	++++	++++	++	++

[a] Cytopathogenic effects evaluated by examination of 3 microtiter wells. Number represents the mean of three experiments. Individual wells were scored either 0 = no cytotoxicity, ± = 10%, + = 25%, ++ = 50%, +++ = 75%, or ++++ = 100% cytopathology. HSV added to confluent cultures of MRC-5 (human fibroblasts) on day 0.
[b] Final concentration LiCl added to HSV-infected monolayers.
[c] Plaque-forming units of virus added to initiate the infection.

only). Lithium chloride at 60 mM was cytotoxic causing overwhelming cell death in 48 hours, as determined by trypan blue dye exclusion. Lower concentrations of LiCl were not toxic when cells were cultured in the presence of LiCl for as long as seven days. Data in Tables 12.1 and 12.2 demonstrate that LiCl clearly inhibited, but did not preclude the development of cytopathology associated with viral replication during the seven-day period. The ability of LiCl to inhibit the development of virus-associated cytopathology was more pronounced with varicella-zoster virus infection in vitro. Additionally, 20 to 30 mM LiCl added to cultures resulted in a reduction of HSV yield up to two $\log_{10}$ at 7 days postinfection (15) (Table 12.3).

Would replenishment of LiCl in virus-infected cell cultures affect the yield of virus? This question was approached experimentally by supplementing HSV-infected cultures with additional LiCl on successive days (Table 12.4). The data indicate that replenishment of lithium to the HSV-infected cell cultures prolonged

TABLE 12.2. Inhibition of the development of varicella-zoster virus associated cytopathology in vitro by lithium chloride.

Virus added	Day post infection	Viral cytopathogenic effects[a] Lithium chloride concentration (mM)			
		0	10	20	40[b]
2×10^2	2	+	0	0	0
PFU	3	+	±	0	0
	7	+++	++	0	0

[a] Cytopathogenic effects evaluated by examination of 3 microtiter wells in each of two experiments. VZV added to confluent cultures of human foreskin fibroblasts on day 0. CPE scored as in Table 12.1.
[b] LiCl added to monolayers at the same time as virus; final concentration in cultures is indicated.

TABLE 12.3. Effect of ethylenediaminetetraacetic acid (EDTA) on lithium associated inhibition of herpes simplex virus replication.

	Virus titer (plaque forming units)[a]	
	−EDTA	+EDTA (0.1mM)
Medium only	35,500	40,500
LiCl 10 mM	2,500	1,130
20	225	135
30	167	155

[a] Rabbit skin fibroblasts were simultaneously infected with HSV and treated with LiCl ± EDTA for a period of seven days. At this time cells and supernatant fluids from infected cultures were collected and quantitated for virus in a plaque-forming assay.

the antiviral effect as reflected by the retardation in the development of viral cytopathology, in many but not all cases. Thus, the slow accumulation of lithium over several days to levels that would be cytotoxic as a single dose seems to be more effective. The reasons for this observation are not yet clear but may be related to maintaining a higher level of available lithium ions.

When the kinetics of lithium addition to HSV-infected cultures was tested a decrease in viral titer was noted when LiCl was added at 2 but not when added 6 hours postinfection (unpublished observations). Thus, lithium may exert an effect on early events in the cycle of viral replication. The literature contains these limited observations and the single clinical trial reported by Skinner and as

TABLE 12.4. Prolonged inhibition of herpes simplex virus associated cytopathology by daily addition of lithium chloride.

HSV added	Day post infection	LiCl (mM)[b]	Viral cytopathogenic effect[a] LiCl (mM) added daily[c]		
			0	5	10
10³ PFU	2	0	±	−	−
		10	±	±	±
		20	±	±	±
		30	±	±	±
	3	0	+ +	−	−
		10	+ +	+ +	±
		30	+ +	+ +	+ +
	7	0	+ + +	−	−
		20	+ + +	+ +	+
		30	+ + +	+ +	+ +

[a] Cytopathogenic effects were evaluated by examination of 3 microtiter wells in each of two experiments. HSV was added to confluent cultures of rabbit skin fibroblasts on day 0. Cultures were scored as in Table 12.1.
[b] LiCl added to monolayers at the same time as virus; final concentration in cultures is indicated, before daily addition.
[c] Without removal of any medium, LiCl in 10 μl was added daily to each culture for 6 days.

TABLE 12.5. Effect of magnesium chloride on inhibition of herpes simplex virus associated cytopathology by lithium chloride.

HSV added	Day post infection	LiCl (mM)[b]	Cytopathogenic effect[a] MgCl (mM)[c]		
			0	20	30
10² PFU	1	0	+	+	+
		20	±	±	ND
		30	±	ND	±
	3	0	+ + +	+ +	+ +
		20	+	+ +	ND
		30	+	ND	+
	7	0	+ + + +	+ + + +	+ + + +
		20	+ + +	+ + + +	ND
		30	+ +	ND	+ + +

[a] Cytopathogenic effects evaluated by examination of 3 microtiter wells in each of two experiments. HSV added to confluent cultures of rabbit skin fibroblasts on day 0. CPE scored as in Table 12.1.
[b] LiCl added to monolayers at the same time as virus; final concentration in cultures is indicated.
[c] Concentration of MgCl added at same time as LiCl.
ND = Not done.

yet no in-depth studies to explore the potential antiviral activity of lithium against herpesviruses or other viruses. A major reason for a lack of clinical studies may be the risk of toxicity following systemic administration of this element. However, topical application of lithium in an ointment poses little threat of toxicity and seems to warrant further evaluation. Such therapy would be applicable to exanthematous infections like HSV and herpes zoster. There is currently no satisfactory treatment for recurrent herpes infections, although the administration of oral acyclovir (Zovirax) is used for primary HSV and in some circumstances for recurrent infections. In comparison to lithium, Zovirax is relatively costly. Ointments containing lithium potentially offer the relative simplicity of topical application and an inexpensive alternate to acyclovir.

In vitro treatment of lymphocytes from AIDS patients with lithium-stimulated T cells in 15/22 subject (9). No mention was made of any antiviral effects. It may seem unusual that lithium would be considered for the treatment of HIV, which is an RNA virus, based on the studies of Skinner (11); however, HIV, a retrovirus, utilizes a double-stranded DNA intermediate for its replication, and utilizes a DNA polymerase. This polymerase is very different from other viral DNA polymerases, in that it is RNA dependent. It remains to be determined whether lithium will affect this RNA-dependent DNA polymerase (reverse transcriptase) in a similar manner to that described by Skinner et al. (11).

Mechanism(s) of the Antiviral Effects of Lithium

The mechanism(s) by which lithium inhibits viral replication have yet to be elucidated notwithstanding Skinner's suggestion that inhibition of DNA synthesis is involved via dysfunction of viral DNA polymerase (11). Studies by Bach and

TABLE 12.6. Prolonged inhibition of herpes simplex virus associated cytopathology by daily addition of lithium chloride in the presence of EDTA.

HSV added	Day post infection	LiCl (mM)[b]	Viral titers[a] LiCl (mM) added daily[c]		
			0	5	10
10^3	2	0	10^4	ND	ND
PFU		10	10^3	10^2	$<10^1$
		20	10^2	$<10^1$	$<10^1$
		30	10^2	$<10^1$	$<10^1$
	7	0	$>10^4$	ND	ND
		10	$>10^4$	10^3	10^2
		20	10^3	$<10^1$	10^1
		30	10^3	$<10^1$	$<10^1$

[a] Tissue culture infectious doses of HSV evaluated by examination of 10-fold dilutions of material recovered from 3 microtiter wells. Supernatant fluid containing virus from wells that had been treated as indicated were added to confluent cultures of rabbit skin fibroblasts and evaluated after 6 days. Cultures showing no evidence of cytopathology are listed as $<10^1$ TCID; those showing CPE at the highest dilution tested are listed as $>10^4$ TCID/ml.
[b] LiCl and EDTA (0.1 mM) added to monolayers at the same time as virus; final concentration in cultures is indicated, before daily supplement of LiCl.
[c] Without removal of any medium LiCl in 10 μl was added daily to the indicated cultures for up to 7 days.
ND = Not done.

Specter indicate that lithium may compete with divalent cations, such as magnesium (Mg^{++}) (15), which is a required cofactor for enzymatic reactions, including polymerases. This suggestion is based upon the observation that addition of Mg^{++} at concentrations equivalent to lithium inhibits somewhat the ability of lithium to impede HSV replication. Magnesium alone had no effect on HSV replication (Table 12.5). Furthermore, the addition of 0.1 M EDTA, a strong chelating agent for Mg^{++}, enhanced the antiviral effects of LiCl with a 2-fold reduction in the yield of viable virus (Table 12.3). The most notable antiviral effects were seen when cells were initially treated with LiCl and EDTA and then supplemented daily with additional LiCl (Table 12.6). HSV was recovered either 2 or 7 days postinfection and then tested for infectivity by serial 10-fold dilution on cultures of rabbit skin fibroblasts. Cultures initially treated with 30 mM LiCl plus EDTA and then supplemented with 5 or 10 mM LiCl daily yielded no viable virus in cell culture fluids, whereas similarly treated infected cell cultures that did not receive daily lithium chloride supplements yielded greater than 10^3 tissue culture infectious doses (infectious particles) of HSV. Lower initial concentrations of LiCl also inhibited HSV replication but to a lesser degree than the 30 mM concentration. A possible mechanism of action suggested by Bach is that lithium could work by replacing divalent cations, especially magnesium, in one or more pathway(s) involved in viral replication (16).

Other possible mechanism(s) by which lithium inhibits HSV infection in vivo, as mentioned earlier, may be related to its ability to stimulate the immune system (17,18). These effects are especially important in light of the reported significance of cell-mediated immunity (macrophages and lymphocytes) in the control of HSV and other viral infections (19).

Conclusions

The effectiveness of lithium as an antiviral agent remains to be determined in controlled scientific studies. However, a number of reports suggest that lithium has both antiviral and immunostimulatory activity, indicating it has a potential therapeutic use against HSV infections. Anecdotal reports from patients who have used topical application of ointments containing lithium, as well as the study by Skinner (6), indicate that this element may be clinically effective. The data should pique the interest of the scientific community in the antiviral potential for this element. More comprehensive studies of its activity against the medically relevant herpesviruses and HIV would be a necessary prerequisite to clinical trials, if the intimations of past studies are validated during the explorations of the biological activities of this element.

References

1. Horrobin, D.F. (1979). Lithium as a regulator of prostaglandin synthesis. In T. Cooper, N.S. Kline, & M. Schou (Eds.), *Lithium* (pp. 854–880). Amsterdam: Excerpta Medica.
2. Shenkman, L., Borkowsky, W., Holzman, R.S., et al. (1978). Enhancement of lymphocyte and macrophage function in vitro by lithium chloride. *Clin. Immunol. Immunopathol.*, *10*,178–186.
3. Dosch, H.M., Matheson, D.S., Shurman, R.K.B., et al. (1980). Anti-suppressor cell effect of lithium *in vitro* and *in vivo*. In A.H. Rossoff & W.H. Robinson (Eds.), *Lithium effects on granulopoiesis and immune function* (pp. 47–62). New York: Plenum Press.
4. Lyman, G.H., Williams, C.C., & Preston, D. (1980). The use of lithium carbonate to reduce infection and leukopenia during systemic chemotherapy. *N. Engl. J. Med.*, *302*,257–260.
5. Lieb, J. (1979). Remission of recurrent herpes infection during therapy with lithium. *N. Engl. J. Med.*, *301*,942.
6. Skinner, G.R.B. (1983). Lithium ointment for genital herpes. *Lancet*, *2*,288.
7. Horrobin, D.F. (1984). Lithium in the control of herpesvirus infections. In R.O. Bach (Ed.), *Lithium: Current applications in science, medicine, and technology* (pp. 397–406). New York: John Wiley and Sons.
8. Parks, D.G., Greenway, F.L., & Pack, A.T. (1988). Prevention of recurrent herpes type II infection with lithium carbonate. *Clin. Res.*, *36*,147A (abstract).
9. Parenti, D.M., Simon, G.L., Scheib, R.G., et al. (1986). The effect of lithium carbonate in HIV-infected patients with immune dysfunction. In *Proceedings 2nd Internat. Symp. AIDS*, Paris.
10. Herbert, V., Jacobson, J., Shevchuk, O., Gulle, V., & Sawycky, R. (1988). Lithium in AIDS. *Clin. Res.*, *36*,620A (abstract).
11. Skinner, G.R.B., Hartley, C., Buchan, A., Harper, L., & Gallimore, P. (1980). The effect of lithium chloride on the replication of Herpes simplex virus. *Med. Microbiol. Immunol.*, *168*,139–148.

12. Hartley, C.E. (1983). The effect of lithium on Herpes simplex virus replication. *Med. Lab. Sci.*, *40*,406–408.
13. Specter, S., Bach, R.O., & Green, C. (1986). Inhibition of herpes virus in cell culture by lithium ions. *Abstr. IX Internat. Cong. Infect. Parasit. Dis.*, p. 417.
14. Specter, S., & Bach, R.O. (1987). Lithium ion induced inhibition of herpesviruses in cell culture. *Abstr. Am. Soc. Microbiol.*, p. 16.
15. Bach, R.O., & Specter, S. (1987). Antiviral activity of the lithium ion with adjuvant agents. In N.J. Birch (Ed.), *Lithium: Inorganic pharmacology and psychiatric use* (pp. 91–92). Oxford: IRL Press Limited.
16. Bach, R.O. (1987). Lithium and viruses. *Med. Hypotheses*, *23*,157–170.
17. Lieb, J. (1981). Immunopotentiation and inhibition of herpes virus activation during therapy with lithium carbonate. *Med. Hypotheses*, *7*,891–905.
18. Shenkman, L., Borkowsky, W., & Shopsin, B. (1980). Lithium as an immunologic adjuvant. *Med. Hypotheses*, *6*,1–6.
19. Aurelian L. (1988). Herpes simplex. In S. Specter, M. Bendinelli, & H. Friedman (Eds.), *Viral induced immunosuppression* (pp. 73–99). New York: Plenum Press.

13
Lithium and Dermatological Disorders

David. F. Horrobin

Introduction

This review covers two quite different aspects of the effects of lithium on the skin. First, there are many reports in the literature of the precipitation or exacerbation of dermatological disorders following oral administration of lithium for bipolar affective disorders. The review considers the evidence for these adverse effects.

Second, and more positively, there are increasing reports of the desirable anti-inflammatory actions of lithium applied topically to the skin surface. This evidence is also reviewed. Attempts are made to explain how the same substance might activate skin disorders when administered orally, and inhibit them when applied topically. Understanding of these divergent clinical actions on the skin may help to illuminate the actions of lithium in manic-depression.

Oral Lithium and Skin Disorders

It has been estimated that in sophisticated societies, of the order of one person in a thousand is being treated with lithium for manic-depression (1). This means that in Western Europe and North American at least half a million people are taking lithium at any one time.

Dermatological disorders are common in the general population. The various forms of eczema affect 3% to 4% of the population, while 1% to 3% suffer from psoriasis to some degree. Acne is near universal in adolescent males and affects up to 5% of young and middle-aged adults. Severe seborrhoeic dermatitis affects about 1% of adults, while the minor forms associated with dandruff may affect 10% to 20% of the population. These estimated prevalence figures may be found in any of the major European or American textbooks of dermatology (2,3).

Given the high prevalence of dermatological disorders in the general population, unless for some reason manic-depression protects against skin disease, it is certain that skin disorders will be common in patients taking lithium. Given also that skin diseases are notorious for the way in which they wax and wane apparently spontaneously, it is also certain that such waxing and waning will occur in some patients coincidentally with starting or stopping lithium therapy. The known epidemiology of skin diseases therefore guarantees that there will be many examples of an association between the presence of lithium treatment and the development or exacerbation of a dermatological illness. It is instructive to make some simple calculations as to how many patients on lithium are likely to experience the various dermatological disorders. Suppose that in populations where events are likely to be reported in the literature there are 500,000 people taking lithium. Suppose also that 3% of those populations suffer from some form of eczema, 2% have psoriasis, 5% have acne, and 1% have severe seborrhoeic dermatitis. Then at any one time we would expect 15,000 patients taking lithium to have eczema, 10,000 to have psoriasis, 25,000 to have acne, and 5,000 to have severe seborrhoeic dermatitis. In this situation one must be exceptionally careful about attributing a causal relationship to an observed association between lithium and a dermatological disorder.

Attribution of a causal relationship between lithium and a dermatological disorder requires the following:

1. A prospective study should show that the disease in question is more common in a population of lithium-treated patients with manic-depression than in manic-depressives who are untreated or who are treated in some other way. It is not sufficient to compare lithium-treated manic-depressives with people who are neither being lithium-treated nor are manic-depressive: this will not eliminate the possibility that it is manic-depression rather than lithium treatment which is associated with the disease in question.
2. There should be repeated case reports where the skin disease has been precipitated by lithium, remitted on cessation of lithium, and precipitated again on reintroduction of lithium. Because skin diseases are so common, and because they so frequently show relapses and remissions in the absence of lithium treatment, situations in which initiation or exacerbation of a skin disease is associated with starting lithium, and recovery is associated with stopping lithium are unlikely to be particularly rare. Given the sizes of the populations involved, many such events must occur purely by chance. The critical test should be the effect of reintroduction of lithium.
3. Ideally there should be a plausible mechanism which will explain the effects of lithium.

If these criteria are strictly applied, then there is no solid evidence of a causative association between oral lithium treatment and any skin disease. Given the numbers of people who, on the basis of the epidemiology, must have skin disorders while they are being lithium-treated, the paucity of reports in the literature, rather than their frequency, is what is surprising. The majority of reports relate to maculopapular eruptions, acne, and psoriasis although there are one or two case reports of a number of other conditions including dermatitis herpetiformis and exfoliative dermatitis (4,5,6).

There is unfortunately no reliable information concerning the incidence or prevalence of dermatological disorders in untreated patients with manic-depression as compared to normal individuals. There appears to be only one oft-quoted study which looked at dermatological problems in psychiatric patients prior to the development of lithium therapy (7). All 13,468 psychotic patients in nine psychiatric hospitals were supposedly examined. Apart from the obvious enormity of the task, it must be doubted whether all patients were examined with equal diligence since the prevalence of dermatoses of all varieties varied between hospitals from 0.4% to 3.4%. Such an eightfold variation is inherently unlikely. All that can be said on the basis of this study is that there were no truly dramatic differences from normal in the dermatological conditions found in psychotic patients.

The only study in the literature which was even partially controlled appears to be that of Sarantidis and Waters (5). Ninety-one patients attending a lithium clinic were compared with 44 depressed or anxious patients being treated with minor tranquilizers or tricyclic antidepressants. In the males there were no differences in the skin disorders found in the two groups. In contrast, in the females in the lithium group 4% had a psoriatic disorder, 15% had an acneiform disorder, and 29% had some other form of skin eruption which appeared to be temporally associated with lithium treatment. The corresponding numbers in the anxious and depressed patients were psoriasis 0%, acneiform eruption 4%, and other skin disorder 11%.

This study does not in any way address the question of whether patients with manic-depression may show a pattern of skin disease different from those with other psychiatric disorders. The excess of skin problems in the lithium-treated females could either have been due to the lithium, or to the underlying disorder. The temporal patterns of onset or exacerbation of the skin conditions suggested that lithium might be involved, but in the absence of a true prospective study and testing by withdrawal and subsequent challenge, there can be no certainty about this. The most interesting feature of this study is certainly the difference between males and females. The males showed no evidence at all of any lithium-related skin disorders. The females did show such evidence. The authors considered and rejected the hypotheses that the excess in females might be related to greater skin consciousness, greater suggestibility, or a greater frequency of thyroid problems.

The various skin conditions that have been described in association with lithium therapy will now be reviewed in turn.

Nonspecific Maculopapular Rashes and Folliculitis

Nonspecific dermatoses of these types are common reactions to drugs. They may occur as a result of some hypersensitivity of the patient to the drug or as a consequence of a true allergic reaction involving sensitization of the immune system. They may also occur in response not to the drug itself but to the excipients associated with its presentation.

There appear to be in the literature less than twenty described cases of either maculopapular eruptions (4,5,8,9) or of folliculitis (5,10). In many of the cases the rash resolved either on reduction of the dose or on continuation of lithium treatment. There appears to be only one report where repeated administration and the use of different formulations of lithium has quite specifically linked the reaction to the lithium (11).

Given the very large numbers of patients who have been and are being treated with lithium, there are very few case reports of this type. It can be concluded that there is minimal risk of these types of skin reactions and that the specific association with lithium of some of those which have been reported is uncertain. When reactions do occur, they are likely to be the result of individual idiosyncrasy: it is therefore unlikely that they will throw any light on the mechanisms of action of lithium in general terms.

Acneiform Eruptions

There is more evidence that some forms of acne may be precipitated by lithium. Again, however, given how common acne is in the general population, case reports must be treated with great caution. The Sarantidis and Waters study suggested that acneiform eruptions in females but not in males might be about three times more common in lithium-treated manic-depressives than in patients with anxiety or depression not treated with lithium (5). While there are several case reports in the literature, it is probable that less than thirty such patients in all have been described in published reports (4,5,12–15). It is interesting in view of the Sarantidis study that the majority of other reports of acne-like lesions relate to females. Only one report describes resolution of the lesions in a patient on stopping lithium and a recurrence on rechallenge, establishing that lithium was very likely the cause (14). The same authors found in another patient that the acneiform lesions accumulated lithium, the concentration of lithium in them being about ten times higher than in surrounding normal skin (14).

One of the more convincing pieces of evidence suggesting that in some patients lithium can cause these lesions is that their features enable them to be distinguished from common acne (6). The lithium-associated lesions occur often on the forearms and legs, areas not usually affected in acne vulgaris. Moreover, the lithium-associated lesions were much more uniform than in acne vulgaris and did not evolve in cycles, nor were there comedones or cystic lesions.

Again there seems to be a probable causal relationship in a few patients, especially females. Again also, compared with the very large numbers of patients on lithium, and the frequency of acne in the general population, the effect does not seem to be a major one.

Psoriasis

In the literature on lithium and the skin, most attention has been paid to the possible association between lithium and psoriasis. There are several papers on this but again the numbers appear rather small, less than fifty reported cases in all (4,5,14–19). Some of the cases relate to worsening or precipitation of an attack in patients with a known history of psoriasis, while others describe cases occurring de novo in the absence of any previous history. In only two cases is there a clear description of improvement on stopping lithium followed by worsening on rechallenge (19). In many of the cases described the apparent relation to lithium therapy is weak, with no clear temporal association between starting lithium and precipitation of the condition. Many of the cases therefore probably relate to the normal apparently random fluctuations of the severity of the skin condition in psoriasis.

Yet again, given the probability that at any one time there are around 10,000 patients with psoriasis who are taking lithium, and given the fact that lithium has now been used for about 35 years, less than fifty reported cases, many of doubtful value, does not seem to be a very large number. Particular attention has probably been paid to the issue of lithium and psoriasis because, unlike the situation with the other skin disorders, this time there are at least two plausible mechanisms whereby lithium might initiate or exacerbate psoriatic lesions. One has a suspicion that one reason why the association between lithium and psoriasis is so well known is that it can be used to bolster the arguments for one or other mechanism of the underlying disease process.

The biochemical basis of psoriasis is unknown, but increased neutrophil migration into the skin, and increased activation of the neutrophils that are there appear to be involved, especially in the pustular form (20). Lithium is well known to be able to stimulate neutrophil proliferation in various situations, possibly by blocking an inhibitory effect of prostaglandins on the action of colony-stimulating factors (see Chapters 6 and 11). Moreover, lithium may have a direct effect in stimulating neutrophil degranulation (21). If neutrophils in the skin are one component in the activation of psoriasis it is therefore plausible to see how lithium might exacerbate the condition.

Another concept of psoriasis postulates that low levels of cyclic AMP may be important (22–24). Since lithium inhibits the effects of prostaglandin E_1 and of several hormones on adenyl cyclase (see Chapter 11), lithium might be expected to lower cyclic AMP concentrations in the skin. If low cyclic AMP levels are important, then lithium would be expected to precipitate psoriasis by this mechanism.

Yet another possibility relates to the fact that in psoriasis there are large amounts of free arachidonic acid in the skin, as though normal regulation of mobilization of arachidonic acid had broken down (25). Prostaglandin E_1 may regulate the mobilization of arachidonic acid by inhibiting phospholipase A_2 by a cyclic AMP-dependent mechanism (see Chapter 11). Inhibition of PGE_1 biosynthesis by lithium might therefore exacerbate the excessive arachidonic acid mobilization. Local injections of PGE_1 into psoriatic lesions have been found to promote healing (26).

On balance, although the case is far from solidly proven, it seems likely that in a few patients the provocation of psoriasis by lithium may be real and more than the simple noncausal association between two common phenomena. Even so, given how common both lithium treatment and psoriasis are, the effect is relatively rare, is absent in many and probably most patients, and even when present may be preferable in most to the manifestations of manic-depression. Only rarely is the action of lithium likely to be severe enough to require cessation of lithium therapy.

Positive Dermatological Effects of Oral Lithium

In contrast to the earlier-mentioned possible adverse effects of lithium therapy, four skin diseases have been reported to improve with lithium. These are seborrhoeic dermatitis, contact eczema, atopic eczema, and herpes viral infections.

The first of these illustrates the pitfalls of case reports. Two patients with seborrhoeic dermatitis were reported to improve dramatically on being treated with oral lithium (27). But a placebo-controlled trial of oral lithium in patients with seborrhoeic dermatitis showed no effect, with the lithium group actually doing nonsignificantly worse than the control group (28). Thus a formal trial gave no support to the observations made in the case reports.

In both contact eczema and atopic eczema, a controlled but not blinded study from the Soviet Union has suggested a beneficial effect of oral lithium carbonate (29). This deserves further exploration with an improved clinical trial design.

Finally, in patients with oral and genital herpes, lithium has been reported to produce an improvement in the lesions and cessation or attenuation of recurrent attacks (30,31). These case reports have not yet been followed by a placebo-controlled study of oral lithium, but such a study of topical lithium has confirmed a beneficial effect (see later).

Topical Lithium and Skin Diseases

The effects of topical lithium on the skin may have been noticed inadvertently for centuries. Almost all spa waters which are famous for their effects on the skin are relatively rich in lithium ions. While it would obviously be implausible to attribute all the beneficial effects of such complex compositions to a single ion, it is probable that lithium makes some contributions to the beneficial results. The first explicit reference to a beneficial effect of topical lithium is that of Sherwin (32). Sherwin was the chief pharmacist at the Douglas Hospital in Montreal, the psychiatric hospital where Lehmann introduced lithium treatment to North America in the 1950's. There were no standard lithium preparations available, and Sherwin himself made up all of Lehmann's formulations. Sherwin suffered intermittently from a dermatitis of the hands of obscure origin: he noticed that this resolved rapidly on exposure to various lithium salts. Sherwin prepared several formulations for topical use which were employed until his retirement by staff at the Douglas Hospital for various types of dermatitis and pruritus. But unfortunately Sherwin, in spite of numerous efforts, could never obtain the backing to do formal trials.

Then Lieb, in the 1970's, noticed beneficial cutaneous effects of oral lithium on herpes infections (30,31). Lieb, together with myself and Sherwin, formulated an ointment containing lithium succinate which was tested in various skin diseases on a pilot scale. The effect of the preparation on the lesions of both oral and genital herpes seemed striking, with rapid relief of pain and irritation, and shortening of the attack when the ointment was applied early on. Quite independently,

Skinner in Birmingham, England, had noted that lithium was able to inhibit the replication of herpes and other DNA viruses, while having no effect on RNA viruses (33). Skinner therefore performed a double blind trial of the ointment devised by Horrobin et al. in patients with genital herpes. As compared to placebo, the lithium-containing ointment produced a significant reduction in pain, discomfort, and viral shedding (34).

Seborrhoeic Dermatitis

To date, the most striking effects of topical lithium have been noted in seborrhoeic dermatitis (SD). SD is a dermatitis primarily affecting the sebum-producing areas of the body, including the face and scalp. It is characterized by the production of greasy yellow scales which flake off to leave a pinkish base before reforming. Dandruff and seborrhoeic dermatitis form a disease continuum and many dermatologists believe that dandruff is simply a mild form of SD. The pathogenesis is uncertain, but the lesions contain abundant quantities of a fungus, Pityrosporum ovale, which is also present on normal skin but in much smaller amounts. It is uncertain whether the proliferation of the fungus is a primary event causing the disease, or whether the fungus multiplies as a consequence of some other primary breakdown of the skin defenses. Either way, the fungus contributes to the disease because substantial healing follows treatment with specific antifungal agents such as ketoconazole.

Following a pilot study which showed promising effects, topical lithium succinate or placebo were administered to 17 patients with SD in a double-blind, placebo-controlled, crossover trial (35). Half the patients were randomized to receive placebo first and half lithium first for a period of four weeks. Then patients crossed over to the other treatment. The results were dramatic. All the parameters assessed by both doctors and patients (redness, scaling, area involved, itching, and overall impression) improved highly significantly during lithium treatment as compared to placebo treatment. No effect of lithium in vitro on the proliferation of Pityrosporum could be demonstrated (35). However, subsequently, the same investigators (Burton and colleagues) have found that lithium treatment greatly reduces the amount of Pityrosporum which can be recovered from the skin (personal communication). Thus lithium may in some way restore the skin's immunological or other defenses against the organism.

A much larger study in which over 100 patients were studied in four centers has now produced almost identical results (36). It is therefore clear that topical lithium is set to become a useful addition to dermatological treatment.

Interestingly, although no controlled studies have yet been performed, topical application of lithium ointment to the lesions of psoriasis and acne has not been associated with any exacerbation. Indeed, the preliminary observations suggest that the effect of the topical lithium may be healing of the lesions rather than any worsening. Thus topical and oral lithium appear to have radically different actions on the skin. The explanation is as yet uncertain but may relate to the effects of the different concentrations of lithium on essential fatty acid and eicosanoid formation (see Chapter 11). With oral lithium therapy, the concentration of lithium in the skin will normally be less than 1 mM. With topical therapy the concentration is likely to be in excess of 50 mM. Using very sensitive mass spectrometric techniques as well as the more usual flame photometry, we have not been able to detect any change in plasma lithium concentrations following several weeks of topical lithium therapy.

At low concentrations of lithium in the 1 mM range, there is a relatively specific inhibition of the synthesis of PGE_1. Thromboxane formation may also be reduced, but there is no inhibitory effect on the synthesis of other prostaglandins derived from arachidonic acid. The relatively specific inhibition of PGE_1 production is likely to provoke inflammation: PGE_1 exerts anti-inflammatory effects by inhibiting phospholipase A_2 and so reducing the mobilization of arachidonic acid. Low levels of PGE_1 may perhaps allow greater release of arachidonic acid which would be expected to be pro-inflammatory. However, at concentrations above 50 mM, lithium has a quite different effect. It reduces rather drastically the formation of both 1 series and 2 series prostaglandins, thus blocking the formation of both anti- and pro-inflammatory eicosanoids. It is likely to be the latter effect which predominates when topical lithium is used thus, in part at least, explaining the anti-inflammatory actions of topical application.

Conclusions

Although lithium has a poor reputation in relation to skin disorders, the available data suggest that adverse events are rather rare in relation to the widespread use of lithium. It is probable that very occasionally lithium precipitates or exacerbates psoriasis and acne but only rarely will such a side effect require cessation of lithium therapy.

Exploration of the possible positive dermatological uses of lithium is in its infancy. The effects of lithium in genital herpes and seborrhoeic dermatitis may be indicators of a future range of desirable lithium actions.

References

1. McCreadie, R.G. (1987). The economics of lithium therapy. In F.N. Johnson (Ed.), *Depression and mania. Modern lithium therapy.* Oxford: IRL Press.
2. Rook, A., Wilkinson, D.S., Ebling, F.J.G., et al. (1986). *Textbook of dermatology*, 4th ed. Oxford: Blackwell Scientific Publications.
3. Fitzpatrick, T.B., Eisen, A.Z., Wolff, K., et al. (1987). *Dermatology in general medicine*, 3rd ed. New York: McGraw-Hill.
4. Deandrea, D., Walker, N., Mehlmauer, M., et al. (1982). Dermatological reactions to lithium: A critical review of the literature. *J. Clin. Psychopharmacol.*, 2,199–204.
5. Sarantidis, D., & Waters, B. (1983). A review and controlled study of cutaneous conditions associated with lithium carbonate. *Br. J. Psychiatry*, 143,42–50.
6. Heng, M.C.Y. (1982). Cutaneous manifestations of lithium toxicity. *Br. J. Dermatol.*, 106,107–109.

7. Slorach, J. (1953). Emotional factors in skin disease. In E. Wittkower & B. Russell (Eds.), *Skin and psychoses*. London: Cassell.

8. Callaway, C.L., Hendrie, H.C., & Luby, E.D. (1968). Cutaneous conditions observed in patients during treatment with lithium. *Am. J. Psychiatry, 124,*1124–1125.

9. O'Connell, R.A. (1971). Lithium's site of action. Clues from side effects. *Compr. Psychiatry, 12,*224–229.

10. Rifkin, A., Kurtin, S.B., Quitkin, F., et al. (1973). Lithium-induced folliculitis. *Am. J. Psychiatry, 130,*1018–1019.

11. Meinhold, J.M., West, D., Gurwich, E., et al. (1980). Cutaneous reaction to lithium carbonate: A case report. *J. Clin. Psychiatry, 41,*395–396.

12. Kusumi, Y. (1971). A cutaneous side effect of lithium: Report of two cases. *Dis. Nerv. Syst., 32,*853–854.

13. Yoder, F.W. (1975). Acneiform eruption due to lithium carbonate. *Arch. Dermatol., 111,*396–397.

14. Reiffers, J., & Dick, P. (1977). Cutaneous side effects of treatment with lithium. *Dermatologica, 155,*155–163.

15. Okrasinki, H. (1977). Lithium acne. *Dermatologica, 154,*251–253.

16. Carter, T.N. (1972). The relationship of lithium carbonate to psoriasis. *Psychosomatics, 13,*325–327.

17. Bakker, J.B., & Pepplinkhuizen, L. (1976). More about the relationship of lithium to psoriasis. *Psychosomatics, 17,*143–146.

18. Skott, A., Mobacken, H., & Starmark, J.E. (1977). Exacerbation of psoriasis during lithium treatment. *Br. J. Dermatol., 96,*445–448.

19. Skoven, I., & Thormann, J. (1979). Lithium compound treatment and psoriasis. *Arch. Dermatol., 115,*1185–1187.

20. Christophers, E., & Krueger, G.G. (1987). Psoriasis. In T.B. Fitzpatrick, A.Z. Eisen, K. Wolff, et al. (Eds.), *Dermatology in general medicine*, 3rd ed. New York: McGraw-Hill.

21. O'Riordan, J.W., Kelleher, D., Williams, Y., et al. (1986). Effect of lithium therapy on inflammatory response. *Inflammation,10,*49–57.

22. Voorhees, J., Dwell, E., & Arbor, A. (1971). Psoriasis as a possible defect of the adenylcyclase-cyclic AMP cascade. *Arch. Dermatol., 104,*352–358.

23. Abel, E.A., DiCicco, L.M., Orenberg, E.K., et al. (1986). Drugs in exacerbation of psoriasis. *J. Am. Acad. Dermatol., 15,*1007–1022.

24. Voorhees, J.J. (1982). Cycle adenosine monophosphate regulation of normal and psoriatic epidermis. *Arch. Dermatol., 118,*862–874.

25. Kragballe, K., & Voorhees, J.J. (1985). Arachidonic acid and leukotrienes in clinical dermatology. *Curr. Probl. Dermatol., 13,*1–10.

26. Jacobs, K.F., & Jacobs, M.M. (1974). Prostaglandin treatment of psoriatic skin. *Rocky Mt. Med. J., 71,*507–510.

27. Christodoulu, G.N., & Vareltzides, A.G. (1978). Positive side effects of lithium? *Am. J. Psychiatry, 135,*1249.

28. Christodoulu, G.N., Georgala, S., Varaltzides, A.G., et al. (1983). Lithium in seborrhoeic dermatitis. *Psychiatric J. Univ. Ottawa, 8,*27–29.

29. Kachuk, M.V., & Yagovdik, N.Z. (1987). Lithium carbonate in the treatment of patients with atopic dermatitis and eczema. *Med. Sci. Res., 15,*1423–1424.

30. Lieb, J. (1979). Remission of recurrent herpes infection during therapy with lithium. *N. Engl. J. Med., 301,*942.

31. Lieb, J. (1981). Immunopotentiation and inhibition of herpes virus activation during therapy with lithium carbonate. *Med. Hypotheses, 7,*885–890.

32. Sherwin, L. (1974). Compositions containing lithium succinate. *US Patent*, *3,639,625*.
33. Skinner, G.R.B., Hartley, C.E., Buchan, A., et al. (1980). The effect of lithium chloride on the replication and macromolecular synthesis of herpes simplex virus. *Med. Microbiol. Immunol.*, *168*,139–148.
34. Skinner, G.R.B. (1983). Lithium ointment for genital herpes. *Lancet*, *2*,288.
35. Boyle, J., Burton, J.L., & Faergemann, J. (1986). Use of topical lithium succinate for seborrhoeic dermatitis. *Br. Med. J.*, *292*,28.
36. Gould, D., Strong, A., Moss, et al. *A multicentre trial of lithium succinate in seborrhoeic dermatitis*. In preparation.

Index

DATE DUE

NOV 2 3 1991			
DEC 1 9 1993			
DEC 0 3 1994			
DEC 0 8 1999			